MADNESS ON THE BRINK OF ECO-APOCALYPSE

"Leutjen's book *Madness on the Brink of Eco-Apocalypse* is more than a wakeup call on the climate crisis. It's a five-alarm fire bell telling us that our house, the planet, is on fire. She brings our attention to the fact that, rather than our government helping to put out the flames, they are handing corporations the matches and fuel to keep the blaze burning.

Leutjen's plea is to the average person who is going down on the *Titanic* with her. Sure, some billionaire folks might see themselves heading to moon or Mars colonies, but the rest of us are going down with the ship. We've already hit the iceberg and it's only a matter of time. Leutjen wants us to be awake and alert as we face this disaster, to understand what's been done to us and our planet and our part in our global demise. Not for the faint of heart."

Davina Kotulski, Ph.D., clinical psychologist, award-winning author, life coach, speaker

"I can almost hear Andy Rooney reading to me. His mix of wit, sarcasm and reality was on par with the author's."

Jennifer Weissmuller, environmental law attorney, Pasadena, CA

MADNESS
ON THE BRINK OF ECO-APOCALYPSE

Furious Facts, Dark Humor & SOS Calls to Action[1]

Cheryl Leutjen
Author of *Love Earth Now*

[1] *Plus superabundant superfluous footnotes*

Copyright © 2024 Cheryl Leutjen. All rights reserved.
Published by: EarthKind Press
Cover Design: Chloë Meyer https://www.chloe-meyer.com/
Layout & Design: Polgarus Studio

Madness on the Brink of Eco-Apocalypse: Furious Facts, Dark Humor, and SOS Calls to Action

Library of Congress Control Number: 2024916563
ISBN: (paperback) 979-8-218-48067-7,
(ebook) 979-8-218-48068-4

BISAC category codes
SCI092000 SCIENCE / Global Warming & Climate Change
SCI080000 SCIENCE / Essays
BIO026000 BIOGRAPHY & AUTOBIOGRAPHY/ Memoirs

For permission requests, please contact the publisher at:
EarthKind Press
1225 Cypress Avenue
Suite 3 #V323
Los Angeles, CA 90065
mailto:editor@EarthKindPress.com

For special orders, quantity sales, and corporate sales, please email the author at Cheryl@CherylLeutjen.com.

Dedicated to all the brave souls who are doing what they can to make the world a more compassionate, respectful, healthy, and sustainable place for all of us.

And I do mean ALL of us. For all of EarthKind.

I see you. I honor you. I thank you.

Caveats, Disclosures and Tw[1]

The rage is real. There will be some c*rsing.

This book was written before the November 2024 Presidential election. As a result, some of the information and references may be out of date. While every effort was made to ensure accuracy at the time of writing, the rapidly changing political landscape means that certain details may no longer reflect the current situation. I believe the core messages remain on point and the calls to action are more urgent than ever.

Some of the awful truths may be depressing, especially if you are already feeling climate grief or eco-anxiety. Skip over anything that feels too hard. Give yourself all the breaks you need. We need our wits about us if we are to embrace the extraordinary challenges of our time.

If you think I'm being overly critical of anyone mentioned in this book, know that I wrote it while looking in the

---

[1] Trigger Warning

mirror. I'm just as much to blame for the mess we are in as anybody else who never shilled for ExxonMobil.

Full Disclosure: Okay, yes, I did work for an oil company for a few years, but I was an environmental *compliance* attorney, and I did my best to keep our clients on the legit path to doing the right thing.

Unless otherwise noted, I am not affiliated with any of the individuals or organizations mentioned herein, nor do I endorse their products or services.

The superabundant footnotes are truly awful, containing my extra-random thoughts and superfluous snark. Ignore as your sanity demands. You'll find the verifiable citations in the References and Resources chapter. All citations, including clickable links, can also be found at https://CherylLeutjen.com/references

Contents

Introduction xi
1. I'm Mad At God 11
2. Tipsy Tightrope: Losing My Marbles 29
3. The Band Played On 37
4. Playlist Track 1: Biodiversity 44
5. Make It Hard: Lessons In Letting Go 54
6. Playlist Track 2: Plastics 58
7. My Heart Is This Bamboo Fork 75
8. Playlist Track 3: Dead Zones 81
9. Cooking Up Therapy: Recipe For Resilience 88
10. Playlist Track 4: Deforestation 95
11. The Hate List 110
12. Playlist Track 5: Freshwater 119
13. My Unbucket List 127
14. Playlist Track 6: Climate Change 138
15. Activism For Aging Ragers 163
16. Playlist Track 7: Ocean Acidification 173
17. From The Standup Stage: So Woke 185
18. Playlist Track 8: Air Pollution 189
19. Eco-Grief Therapy: Step-By-Step Guide 196
20. Playlist Track 9: Ozone Layer 202
21. I Get It: I Ridicule Me, Too 208
22. Playlist Bonus Track 218

23. Crippling Ignorance 229
24. Stench: When Eco-Intentions Turn Rancid 240
25. Ranting In The Street 246
26. Environ-Mental: Reinventing Our Brand 252
27. Guerilla Foraging 268
28. Here To Stay: Enduring Turbulent Times 272
29. Closing Thoughts: Good Grief 280
30. Going Forth 292

Book Club Questions 295
Acknowledgments 297
Note From The Author 299
About The Author 301
Also By Cheryl Leutjen 303
Praise For Love Earth Now 305
References & Resources 307
Release 327

"I don't have to tell you things are bad. Everybody knows things are bad... We know the air is unfit to breathe and our food is unfit to eat, and we sit watching our TVs while some local newscaster tells us that today we had fifteen homicides and sixty-three violent crimes, as if that's the way it's supposed to be . . . I don't want you to riot—I don't want you to write to your congressman because I wouldn't know what to tell you to write. . . I want you to get up now . . . and go to the window. Open it, and stick your head out, and yell:

I'M AS MAD AS HELL, AND I'M NOT GOING TO TAKE THIS ANYMORE!"

Lumet, Sidney, director. *Network*. Metro-Goldwyn-Mayer, 1976. 2 hrs., 1 min.

Introduction

As the news reels of superstorms, firestorms and shitstorms play on and on, the bad news no longer restricted to the nightly news shows,[2] I can't stop thinking about this scene in the old movie, *Network*. I don't recall much about the film except for that scene, so I can't say if it holds up to the high standing I've given it in my memory banks. Not everything does—I used to think tube tops were a flattering fashion choice.

But that scene comes back to me time and again, over all these decades. The intensity and the fury of the newscaster, urging, *commanding*! people to stick their heads out of their windows and shout, "I'm mad as hell and not going to take it anymore!"

I so relate to his frustration. His wrath. What I do not understand is why we aren't all shouting out our windows, as we witness ever more ruin of our stable climate, fresh water, soils, oceans, forests, coral reefs, biodiversity, and[3], dare I say, basic civility. All the while some so-called news outlets ridicule peer-reviewed

---

[2] It's true, kids. Back in my day, news broadcasts were shown before and after prime-time evening programming. News reports weren't a 24/7 thing.
[3] ...bees, frogs, gorillas, rhinos, box turtles, orangutans, tigers, lemurs, and I have to stop there, or I'll be too sad to write any more.

scientific studies and armchair "experts" send death threats to hard-working scientists[4] for daring, like their predecessor Galileo, to challenge the culturally convenient lies we've been told about the world. While ultra-conservative billionaires fund disinformation campaigns, and so many elected representatives do nothing but rail against the WOKE[5] agenda. The idiotic assertions ("drink bleach to cure COVID!"),[6] the denial of basic truths ("it snowed today so climate change is a hoax!"), and the absurd deflections of responsibility[7] come at me like a freight train full of coked-up monkeys that I swear I did not order, Universe.

There's so much to be mad about these days, isn't there? Despite my propensity to rant about the small-but-maddening minutiae of daily life, I repress most of my anger over these critical issues that threaten Life as We Know It. The problems loom too large and too many—I just cannot get my arms around the magnitude and multitudes. But I know they are deadly, lethal for our kind, dimming the bright futures we want for our children. I'm furious that we've let it come to this.

It's not easy containing the rage. Confining it in a smoldering tinder box requires half of my dwindling energy.

---

[4] Some 40% of 468 climate scientists surveyed in late 2022 say that they've experienced online harassment because of their work; it's 73% for the scientists who post at least once a month.

[5] Anyone else find it ironic that the term "woke," which some use to deny systemic racial injustice, originated from Black culture—as a warning of potential violence against Black People? Can't make this stuff up.

[6] E.g., drinking bleach might cure or prevent COVID-19. Five states reported an increase in calls to poison control centers after Trump's comments about drinking bleach as disinfectant.

[7] Delivered, ninety-nine times out of ten, by the person who is absolutely the most responsible, per my unofficial survey.

But I've given it everything I've got, terrified I'd self-immolate if I unleashed all that pent-up, white-hot fury.

I'm also a deeply spiritual, New Age type person, well-schooled in the Law of Attraction ("LOA" for the disciples), and that's where the real trouble begins. Believe it or not. The LOA, as I understand it, is a fancy way of saying "like attracts like," and that includes our thoughts. If we are thinking about being broke all the time, the Universe will keep sending us more experiences of being broke.[8] So if I want more hope, more happiness, more joy in my life, then I must project more hope! More happiness! More joy!

Sure, that's an oversimplification, but few things please me more than wrapping up something squirmy in a neat package, like swaddling a soft-scented baby that was a poop-covered mess just a bath ago. Chef's kiss. I find fewer and fewer opportunities to do that nowadays, though. The world around me seems a palette of murky hues reminiscent of diarrhea, the black-and-white clarity of my youth vanished like my waistline.

I'm doing my best to think the happy thoughts—which requires a gargantuan effort these days, akin to birthing octuplet elephants—given the poor prognosis for our species on planet Earth. Did you see the news that, not only was 2023 the hottest year on record, but that all ten of the hottest years have occurred in the last decade? That the coverage of Antarctic Sea ice hit a record low in 2023? That the oceans are warming at an

---

[8] The Universe can be a real ass sometimes. PLEASE DON'T SEND ME ANY DONKEYS.

"unprecedented"[9] rate, which means they can absorb less of our excess CO_2? That the collapse of the entire Amazon ecosystem could occur much sooner than we ever imagined? Or that plastic, the "forever waste," is now so pervasive that it's in our lungs, livers, kidneys, breast milk, testicles, and our infants? Or that scientists give us just a handful of years to make massive reductions in emissions or we're toast?[10] How fortunate that all this is unfolding just when the chances of the U.S. tackling these crises post-November 2024 Presidential election are slim to none . . . she said dripping with sarcasm.

And still, I've soldiered on, battling with the ferocity of my cat hopped-up on too much catnip, trying to stay aboard that power-of-positive-thinking train. I can't utter the words "I'm angry" aloud without looking for the lightning bolts. Ridiculous.

And dangerous. I've been forestalling a full-out venting of my rage with the clench of a boa constrictor, but bottling up all the flaming fury I feel isn't good for my mental health. I'm grateful my parents taught me about meditation instead of shooting guns to deal with emo angst. I'm so ill-suited for jail.

So, as of today, with you, dear reader, as my witness, I'm relinquishing my grip and jumping off the cliff into the land of Fury—praying I land with more agility than

---

[9] "Unprecedented" being the most overused word of our times. I caution against making it the trigger word for your drinking game while watching the news.

[10] But, no worries, I heard those rascally scientists who have devoted their lives to serious research can't be trusted; they're just in it for the money. We've all seen how many scientists make the Billionaires List, right?

the time I flew over the handles of the motorbike I had just "learned" to drive.[11]

Today, I am giving myself permission to purge the pent-up negativity like that scene in the horror classic, The Exorcist, where the possessed child spews green projectile vomit. And rant like I've never ranted before. Forget the years of training in Midwest Nice. Say it all without consideration of the karmic blowback—even to the point of revealing my cringiest confession: I'm mad at God. I feel we've been set up to fail. Fail at solving climate change, fail at caring for the biodiversity on which humans rely, fail at protecting the natural resources that support Life as We Know It.

Which makes me even madder.

More on that later.

If you can relate to my fury, in any small way, if the death spiral in which our once-bountiful Nature churns keeps you up at night, this book is for you.

But it's not for everyone. This is not a book to read if you're looking for flowery prose,[12] positive affirmations or simplistic "change your light bulbs and all will be well" platitudes.[13]

I'm not here to convince anyone that climate change is not only real, but deadly for our kind. Or that preserving habitat for red wolves and mountain lions is more important than new mini-malls and mega mansions. Or

---

[11] Explaining the blood and bruises when I went back to work at the movie theater after that lunch break was interesting. I can't recall, but I hope there was a gory film playing that day.

[12] If you ARE looking for flowery prose, I recommend reading my prior book, *Love Earth Now.*

[13] Or if superabundant and superfluous footnotes upset you.

that giving up coal-fired plants and gas-guzzling vehicles is essential if we want a climate that sustains Life as We Know It.

If the science, the hellfire storms, and all the alarm bells ringing aren't enough to convince you that our benevolent, life-sustaining environment is worth protecting, then I suggest you keep your TV tuned to whatever station feeds you the disinformation you like to hear.

"Today, no nation can find lasting security without addressing the climate crisis. We face all kinds of threats in our line of work, but few of them truly deserve to be called existential. The climate crisis does."

Secretary of Defense Lloyd J. Austin III
Leaders Summit on Climate, April 22, 2021

My patience with "this isn't the right time" and the "science is uncertain" crap is long spent, like my tax refund before it even arrives. I mean to act with the urgency that's worthy of the many crises we face as a species on this planet. As a recent cancer survivor, my life expectancy is not what it used to be. I intend to be a force for positive change while I can, as a lasting gift to my children.

If you're even a fraction as concerned as I am, I hope you'll read on. Maybe you'll find some camaraderie in our collective rage. Maybe we will even find some peace through the catharsis. But I make no guarantees. There just aren't any anymore. That's something that those of us who were convinced Hillary Clinton would win the 2016 election have learned.

Witnessing the destruction of our beautiful world, knowing humanity has had so many chances to make more life-affirming choices, creates the perfect recipe for a scalding soup of anger, anguish, and frustration. I mean to steep in that furious stew until my blood stops boiling. The way a 104-degree fever brings on chills. That's my plan, anyway.

Only then will I be any good to anyone, because trying to hold it all in makes me too exhausted to act. To appreciate the beauty that, thanks to a host of miracles, still surrounds me. To pick up my hatchet and bucket so I can continue to chop wood and carry water.[14]

"This is a dark time, filled with suffering and uncertainty. Like living cells in a larger body, it is natural that we feel the trauma of the world. So don't be afraid of the anguish you feel, or the anger or fear, because these responses arise from the depth of your caring and the truth of your interconnectedness with all beings."
Joanna Macy, renowned environmental activist and scholar of Buddhism and deep ecology

---

[14] Taken from a Zen Buddhist saying:
"Before enlightenment, chop wood, carry water.
After enlightenment, chop wood, carry water."

This book is a mish-mash of rants filled with furious facts and rambling essays about my own frantic efforts to cope with the madness, sprinkled with spurts of comedic relief.[15] In other words, it's about as chaotic as the times in which we live. That's just how I roll these days, lurching and reeling, as each new terrifying wave crashes over the bow of my rickety old ship.

I was prepared for some disruptions as we transitioned from the Old Ways, and we figured out how to live sustainably. I just never imagined they would give way to . . . . even older, antiquated ways, the norms circa 1950, say.

I tell myself it's the "last gasp of the old gas,"[16] and that entrenched systems and hierarchies aren't going to go quietly in the night.[17] But it's hard to stomach, like eating Flaming Hot Cheetos when my acid reflux flares up.

If I am to stay well-informed (and the jury is still out whether that's a healthy choice for me), I must digest the furious facts in bite-sized chunks. And then follow up each chunk with one of my many coping and, *dare I say it*, thriving mechanisms that keep me upright and functioning. On most days.

---

[15] That's how bad things have gotten—this introvert has turned to stand-up comedy to get some things off her Earth-loving and plastic-hating chest.
[16] "Gas" being a definitely on purpose pun, given the stranglehold the fossil fuel companies have had on our abilities to change and adapt for so long.
[17] Exhibit A being recent Supreme Court decisions that hamper the efforts of the U.S. Environmental Protection Agency to respond to the climate emergency. As if tackling climate change wasn't hard enough!

I invite you to pick and choose your own mix of chapters to read, tackling the deep dives when you're sufficiently steeled, and opting for the shallow swims when you're not. We've already put off making the necessary systemic changes essential for our survival for decades now. Instead of rushing in like well-meaning missionaries who unwittingly kill their converts with smallpox, let's give ourselves some reflection time. Get some facts. Ponder. Then, leaning on the love in our hearts for all we hold dear, make our considered plans of action. And give them all we've got. This is no time for playing small.

Those action plans may include joining forces with others who share your pain, your vision, your desperate need to be a force for good at this harrowing time. Most chapters in this book end with a showcase of one of the many, many, many groups, clubs, nonprofits, and "cussed stubborn" folks who have channeled their own fears and furies into positive actions.

Perhaps one of their missions will inspire you to form your own posse. Or to seek out an existing chapter near you. Doing something meaningful is the best antidote to despair that I know. Doing it in the company of like-hearted souls amplifies that curative salve a hundredfold—at least it does for me.

We are stronger together. Let the heartless destroyers know that we will not quietly sip our Merlots and whiskeys,[18] while turning a blind eye to the devastation

---

[18] I don't mean to disparage the periodic use of numbing tonics. But they have yet to solve any of my problems, no matter what reassurances they provide in the moment.

that the status quo wreaks.

Be good to yourselves, no matter the speed at which you're navigating the grave challenges of our time. "Self-care" may be the most overused trope today, and yet it's essential if we are all to remain functioning, doing what we can, and, yes, thriving.

Now, take a deep breath, and...

.
.
.
.
.
.
.

Let's get this rant party started.

Exercise our Right to Free Speech as loudly as we can, while we still have it.

What's making YOU mad?

Come meet me at the window.

Bring a megaphone.

GO.

ONE
I'm Mad At God

On any given day, you may find me ranting. No surprise there. Ranting about corporations that knowingly poison the environment, then spend millions on a Superbowl ad to convince us to trust them. Ranting at manufacturers who spew plastic around the planet, like a mob of ravenous raccoons flinging garbage from every can with zero plans to clean it up. Ranting about instruction manual print in a font so minuscule I can't read it with two pairs of reading glasses.

Here's the rant I've been saving up. The one that seemed too awful to speak aloud, saved up for screaming into the void of my closet. Until now.

I'm mad at God.

There, I said it. Waiting for the lightning bolt.

I feel the Divine One has set us up to fail. Fail at solving the climate crisis. Fail at caring for the environment that sustains Life as We Know It. Fail at solving the very problems we humans seem supremely qualified to create.

We know that saturated fats clog our arteries and that heart disease is the leading killer in the U.S. Why can't we stop stuffing our faces with bacon cheeseburgers? I'm not one for conspiracy theories, but something seems hinky here.

So many times, we've been warned that our fossil fuel addiction is heating up the planet, fueling off-the-chart superstorms and firestorms, submerging entire island nations, drying up croplands, proliferating disease-carrying pests, and more. Heck, I studied global warming in college way back in the prehistoric era of the 1980s. Back when it was just considered "science," not a boiling-hot point of political contention. Did you know it was a Republican U.S. President, George H.W. Bush, who signed the United Nations Convention on Climate Change treaty in 1992? And we all know what a left-leaning rabble-rouser he was.[19]

President Bush said, "We must leave this Earth in better condition than we found it, and today this old truth must be applied to new threats facing the resources which sustain us all, the atmosphere and the ocean, the stratosphere and the biosphere. Our village is truly global."

U.S. President George H.W. Bush
Comments at 1992 Earth Summit in Rio de Janeiro

---

[19] That's sarcasm. President George H.W. Bush was known for preferring stability to radical change. And yet he championed amendments to the Clean Air Act to reduce emissions and eliminate acid rain, which passed Congress with "overwhelming bilateral support." Bush also oversaw the passage of the Energy Policy Act of 1992 which included incentives for increased energy efficiency and renewable energy projects.

In 1990, President George H.W. Bush also established the U.S. Global Change Research Program, authorized by Congress, to coordinate federal research on climate change. The vision statement for the program, which still exists today, gives me the sad-chills:

> "A Nation, globally engaged and *guided by science, meeting the challenges of climate and global change* for the benefit of all."

How did we go from the days of this conservative, environmentally-concerned presidency to a time when ridiculing peer-reviewed scientific studies—while continuing to throw subsidies at fossil fuel companies[20]—is the norm for so many politicians? And their approving constituents. And the "news" outlets they follow.

I must have missed a memo somewhere because I can't fathom how our public discourse (or what passes for it these days) has taken such a U-turn. Why are we now, when scientists say we only have until 2030 to make some systems-busting changes, spending more time arguing about *whether* climate change is happening than what to do about it? Something just doesn't add up.

Sure, ExxonMobil has spent millions on disinformation campaigns, designed to cast doubt and dispersions on climate science.[21] And the certain "news"

---

[20] The International Monetary Fund estimates that subsidies to fossil-fuel producers by governments around the world totaled $7 trillion in 2022. If fossil fuels are so great, why do they need all those subsidies?
[21] Check out journalist Amy Westervelt's *Drilled* podcast for a detailed report on ExxonMobil's disinformation campaign. It made my blood boil. If you are similarly prone, I recommend investing in a punching bag before you dive in.

outlets sure do their best to fearmonger. But why have so many of us fallen for the lies and trickery hook, line and sphincter?[22]

---

*"What good is it to save the planet
if humanity suffers?"*[23]

Rex Tillerson
Former Secretary of State and CEO of ExxonMobil

Because if the science doesn't sell, how about looking out of a window? Glaciers are melting like ice cream on an August sidewalk, 100-year storms are arriving more often than Amazon deliveries, and hellscape fires are incinerating areas the size of Massachusetts in the US each year. Why are we, as a species, not united against this existential crisis, doing everything we can to save ourselves? Where's the outcry and rush to take action? A PTA fundraiser generates a bigger flurry of activity than the destruction of the environment that makes life for our kind possible.

My inner toddler, full of rage, says it's God's fault, that we've been set up to fail.

I know. My stomach churns and my heart leaps into my throat to have the gall, as a mere mortal, to blame our

---

[22] See "Our Biggest Fear Is Each Other" in References.
[23] As if climate change isn't already causing plenty of suffering. Ask somebody in Bangladesh. Or watch the movie, *Once You Know.*

Creator. It's taken me years to come to grips with my irreverence. God has Her[24] reasons, I'm sure, though I'm hard-pressed as ever to comprehend them. I realize I risk being smote with pestilence for questioning His Divine Wisdom.

But I keep coming back to this: nothing I'm writing here comes as a shock to the delicate sensibility of the Divine One. Being All Knowing, God is already well aware of everything I'm writing on these pages.

I'm a "better to ask for forgiveness than permission" sort of person, anyway.

So here goes.

I have a few things to get off my chest, God.

I think You've set us up to fail this ultimate test as stewards of this lovely planet. See, I've been reading about all the reasons why our human biology makes us so bad at addressing longer-term threats, and it's not looking good for you, God, I'm sorry to say. I'm no neuroscientist—I studied rocks in college and most days I still prefer them to people—but what I have gleaned from the experts seems to prove my theory: You've set us up to fail.

You've programmed our human brains, across some 2 million years of evolution, to recognize and respond to immediate dangers, like LION! But the slow creep of crepey skin taking over my aging body? Never saw that happening. Broke your leg? Let's rush you to the ER! But

---

[24] Yes, I believe that God contains our multitudes, all ethnicities, all races, all genders, and I choose pronouns accordingly. Feel free to adapt to reflect your own understanding.

if you're slowly dying inside because there's no hope left for our kind? Take two Ambien and don't call me in the morning.

Daniel Gilbert, a social psychologist at Harvard, summed it all up for me best, and I appreciate You sending him my way. He says our human brains excel at detecting and deflecting threats, and I say we should all thank Thee for that. Our species would not have lasted long if we operated like Congress, taking weeks of hearings and senseless debate before deciding that yes, indeed, we should run to avoid that lion.

However, You've set it up so that our brains prioritize certain types of threats, and here's where the trouble starts. Gilbert sums it up this way: our brains prioritize responding to threats that are intentional, immoral, imminent and/or instantaneous. The I's have it! I do adore an apt alliteration.

Let's start at the top, shall we? Things done *intentionally*, like that lion salivating at you as a nice, fat human-steak, get our brains moving a lot faster than, say, the outbreak of a new flu. A virus doesn't act intentionally—it has no brain at all—so it's not "out to get us" the way the lion is. There's just no target to aim our arsenals of deadly weapons at, so how can we expect people to, well, get up in arms about it?

Hollywood gets it. How many movies invent enemies that we can eviscerate with our guns, bombs and missiles? You could learn a thing or two from the Entertainment Czars, God.

Second, anything *immoral* or "indecent"—basically whatever the Moral Police are tsk-tsking about on social media these days—gets a quick response (or knee-jerk reaction?). I guess that explains why burning a book gets more press than someone firing up a greenhouse gas-spewing vehicle. Is conforming to a cultural code more important than preserving our life-sustaining environment, God? I'm starting to wonder about Thy priorities.

The third category consists of threats of *imminent* harm—the shut-off notice from the electric company gets my attention faster than the tax bill due four months from now. Until April 14th rolls around, anyway, then it's all hands on deck.

Caring about the future is such new territory for our brains that the feature is still in beta testing, Gilbert reports. Only a small fraction of our gray matter considers what lies ahead, as compared to the vast mental territory devoted to avoiding baseballs speeding in our direction.

"Our brains are essentially get-out-of-the-way machines constantly surveying the environment trying to find things out of whose way it should right now get. That's why we can duck a baseball in milliseconds."
Daniel Gilbert
Psychology Professor, Harvard University

I guess that explains the bacon cheeseburger dilemma—dying of heart disease is just too far down the road to squelch our "I-want-it-now" appetites. Or caring about the impacts of burning fossil fuels some fifty years down the road, even less. Given our struggles with all manner of addictions to down-the-road-dangerous substances, from opioids to petroleum, could You pretty please speed up the development of this future-focused part of our brains, God? Given the dire circumstances and all.

As for the fourth I, *instantaneous*, the human brain detects changes in all kinds of things, temperatures, sounds, light, but our responses depend on the *rate of change*. Things that show up quickly, like a twister dropping out of the sky, get no end of press. But, like the proverbial frog in a pot of water gradually heating,[25] the slow loss of butterflies and permafrost fails to ring the alarm bells.

Though we're feeling the effects of climate change more and more, Gilbert says it's still unfolding too slowly to trigger an alert to warp speed. You found the sweet spot there, didn't You, God? Change happening too stealthily for our brains to prioritize any response, but too fast to give us time to evolve to solve it. Kudos to You, I guess.

But wait, there's more. Cognitive psychologists talk about our inherent biases that keep us from picking up a firehose to put out the conflagration. The "loss aversion" bias makes us more concerned about losing something in the short-term, like cheap cheeseburgers, than facing what might be even more alarming down the road, like annihilation. The "optimism bias" keeps us overestimating the likelihood of a positive outcome and

---

[25] Not actually true, by the way. Even a frog knows to jump out of a pot of water when it gets uncomfortably warm.

underestimating negative events. Is that why I keep starting new diets, God? *I'm sure this one will work!*

Thanks to our "self-interest bias," we tend to prioritize our own interests over those of others or the greater community. Far be it for the rest of us to concern ourselves with the Pacific Islanders whose homelands are submerging into rising seas.[26] Or the Bangladeshi women continually dredging mud from the river in the futile effort to keep their homeland above the ever-rising floodwaters. Why should we deny ourselves all those Amazon shipments because 7 million Bangladeshis (and counting) have been displaced due to extreme flooding, destroying their homes and croplands, caused by climate change?

"I think the future of humanity in general can go in either direction. We can either go in terms of solidarity or we diverge between the rich and the poor, and between some countries and others. . . It's about sacrificing more and more of the world in order for a few people to continue to live. It will be a fortress world."

Saleemul Huq, Director of the International Centre for Climate Change and Development in Bangladesh.

---

[26] Pacific Islanders contribute 0.03% of global emissions. How is it fair that they lose everything in exchange? That's as infuriating as politicians voting against student loan debt forgiveness as a fiscally irresponsible bailout—but not complaining when their six-figure Paycheck Protection Program loans were forgiven.

Sum all these things up, and it sure sounds like Someone has set us up to fail, if You know who I mean. In my blasphemous book, anyway. I could go on, but it's a steaming, stench-filled heap to process. I'm already on climate grief overload, weeping into my shade-grown coffee every time I do more research. Self-care is something I've had to learn to prioritize—or maybe that's part of the me-first programming, I can't say.

"Global warming is a deadly threat precisely because it sneaks in under the radar that we've evolved. It invades the brain's ancient alarm system and it leaves us asleep in a burning house."
Daniel Gilbert
Psychology Professor, Harvard University

I need a sanity break. Excuse me while I work a mind-numbing crossword puzzle. It feels so good to be able to *solve* something. . . . especially in these techno-stressful times. I can't even figure out how to watch local TV channels, a feat so simple I once accomplished it with two metal sticks called "rabbit ears." But, somehow, I am supposed to figure out how to solve climate change?
BRB.

I'm back. Whew. That's a lot to process, especially now that we're all so deficient in the attention span department. Which is ~~possibly~~ definitely another strike against us. What was I saying?

Being a person who aims to be fair and respectful, I am trying very hard to see both sides, to give You a break here, God. I believe You mean well, and a lot of these priorities and biases make sense, evolutionarily speaking. Immediate problems deserve immediate attention. If I'm too busy planning my party on Saturday night to notice that hungry lion staring me down, well, chances are there won't be any party on Saturday night. Not unless it's my "Celebration of Life," which by the way, I hope is riotous fun with a killer indie-rock playlist, free-flowing kombucha and dancing with wild abandon. I just hope it's not *this* Saturday night.

But what about evolution? "We just need to evolve," right? That's what a well-meaning friend said recently, out of concern for my mental health, I'm sure. I summoned all my will to not bite off his head. Because here's the thing: biological evolution is slow. I can't say how slow exactly, but I recall this much from my geology classes: it's slower-than-sloth slow. Humans have been evolving for some two million years (or thereabouts), and yet we're still focused on scoping out hungry tigers. Meanwhile, social media bots and Tucker Carlson push all our fear-based buttons to convince us that anyone who looks, sounds, thinks, loves, or prays differently than we do is a ravenous tiger.

I am sure You had Your reasons when You set the evolutionary clock at the speed of sloths at the DMV[27],

---

[27] I'm referring to the ones in the movie, *Zootopia*, of course. Real DMV employees put up with a lot, and I mean to attract good service every time I visit. So, Law of Attraction style, I am sending love to the real DMV employees everywhere.

caution being a virtue and all that. But what about our free will—isn't that a characteristic feature of humanity, the ability to choose to rise above our "baser instincts"?

As I was about to hang my hopeful hat on this distinctive ability, along comes the pigeon splat on the crown of my cap. Just today, I learned of neurobiologist Robert Sapolsky's declaration that free will is a myth. Sapolsky studied humans and other primates for forty years before reaching this hope-killing conclusion: we have no more conscious control over our behavior than I have over free books at a yard sale. Sapolsky says our decisions and actions are as voluntary as a muscle spasm, each the inevitable result of our biology, our brain chemistry, and life circumstances. It sure seems like You enjoy pulling the rug out from under all my hopefulness, God.

"We've got no free will.
Stop attributing stuff to us that isn't there."
Stanford neurobiologist Robert Sapolsky

In one respect, I'm deeply relieved. No matter how many SMART[28] goals or New Year's resolutions I set, my decisions to eat Doritos instead of carrot sticks are not my fault.

My other conclusion, however, sends me back to my closet, screaming. Does this mean I can't justifiably rant

---

[28] Corporate speak for goals that are specific, measurable, achievable, relevant and time bound.

about people making horrible decisions because they. . . *just can't help it?*

This free-will-is-a-myth revelation does nothing to improve our chances of addressing long-term threats to our existence, God. Even if everyone in the world suddenly acknowledged the existential threat of a runaway climate, there's nothing any of us can do to make the radical changes required to save our kind? Is that it?

I feel sick. Maybe I should take off this crapped-on cap.

Let's review, shall we? I'm getting lost in the kudzu.[29]

By Your Grace, God, humans evolved on a verdant planet with a most magical atmosphere unlike anywhere in the Universe, so far as we know, perfectly designed to provide for our every need. You've also gifted us, thanks to a whole lot of dead plants, *a moment of silence please*, turned into an abundance of this magical substance called petroleum. Of all the species that arrived before us, only we homo sapiens, with our crafty intelligence, learned to process it into a myriad of miraculous substances, from gasoline to slip-n-slides.

Coupled with these gifts, You saw fit to implant into us humans all those above-mentioned "threat NOW" priorities and "me first" biases against being able to stop the burning of said magical substance in time to save

---

[29] Kudzu, an invasive weed that's native to Japan, outcompetes (i.e., kills off) everything from native grasses to mature trees (by shading them from the sun). A lot like fascism, it's best to root it out before it takes hold . . . and takes over.

ourselves. Those built-in biases render us helpless to fix urgent, modern-world crises, but don't count on evolution to help us fix anything. Not soon enough for us, anyway.

Being a New Age spiritual type, I tell myself, I really try to convince myself, that it's all On Purpose, God, that "everything happens for a reason," and all the spiritual blah blah blah that sometimes gives me comfort. But not so much, *not at all*, when I'm imagining my future grandchildren cursing me for the role I've played in killing off the most beneficent climate, the lushest garden in the Universe.

WTH, God?

Sigh. I'm feeling drained, but nowhere near the relief I expected after a rant of this magnitude. I'd hoped for the kind of vindication that comes from a purge after too many pepperoni pizza Hot Pockets.

Does railing against the One Who Made Us solve any of the problems? No, it does not. Nor do I believe, in my heart of hearts, that it absolves humanity for our failure to employ enough common sense to not poison our own watering holes. Humans take pride in our intelligence as the thing that sets us above and apart from other species, and yet I don't see the other species lighting their homes on fire and calling it God's will. Or essential for "progress."

See, the data has been there for decades. Despite all my ranting about God's part in all this, I have to admit. . . we KNEW. We knew the dangers of our fossil fuel-burning habits back when there was still plenty of time to avoid the worst effects. We know even more now, when there's yet a sliver of a chance we can avoid the very worst effects. And still most of us, in the Western world, at least, seem to lack the collective will to pick up the tools we have been given to save ourselves.

Not that there aren't a whole lot of people doing good things to reduce our carbon emissions, protect habitats, and make renewable energy resources more cost-effective and reliable. On good days, I think of them. I drop to my knees and give thanks for them. Weeping and sobbing.

And I marvel at the indigenous peoples who somehow overcome all these strikes against us to live in harmony with the land. Whenever the rest of us will leave them alone long enough, anyway.

On the not-so-good days, all these efforts seem too little, too late.

And so, I bring it all to God. It's a deal we have. I confess all my frustrations, fury, and ill-placed blame to Him on the floor of my closet. Where it goes from there, well, it gets complicated. I have no intention of explaining my messy belief system here, much less inflicting it on anyone else. There's enough of that going around already. It's my personal faith crisis, one which I'm adding—because why not—to the long list of crises I already feel wholly ill-equipped to navigate.

Suffice to say that we're in talks, me and God, like besties sulking over who said what to whom. Sure, that's an insult to God, bringing Her down to my level. Maybe I take the "image and likeness" thing a little too far, but one thing I believe with a fervor I usually reserve for my first pumpkin spice latte of the season is this: we are all modeled after our Creator, made of the very stuff that God is. Which means that God knows how I feel and loves me anyway. In Her Infinite Benevolence, She shows up for me, blooming the jacarandas along my path, feeding me fresh food, and delivering my loved ones safely home to me at night. I am beyond words grateful.

I also figure the Almighty has run into this kind of resistance to the Divine Plan before. Yet the Benevolent One still cranks out the humans, despite our obstinate and querulous tendencies. God can handle one more cranky old grouch wishing she knew a cheat code to hurry up and advance to the next level. Because she's exhausted, and this level is just too effing hard.

Sorry, God.

"[T]he Divine cannot be described in one color, one tone, one hue; cannot be contained in one image, one metaphor, one symbol; cannot be bound by one tradition, one story, one idea. God is bigger and wider and more than any of our best and worst attempts to name God. . . ."
Mallory Wyckoff

CITIZENS' CLIMATE LOBBY

If you want to connect with folks in your area who are actively working on national issues relating to climate change, check for the nearest chapter of the Citizens' Climate Lobby (CCL). Their approach is nonpartisan, focusing on shared values over partisan divide. What a breath of fresh air.

CCL chapters organize around local congressional districts, currently numbering 534, across all 50 U.S. states, the territories and around the world. If there's not a chapter near you, CCL invites you to contact them anyway, and they will put you in touch with leaders in your area.

"Citizens' Climate Lobby is a nonprofit, nonpartisan, grassroots advocacy climate change organization focused on national policies to address the national and global climate crisis." https://citizensclimatelobby.org/

"One of the main things we do in the greater Dallas/Ft. Worth region is to provide an outlet for people who care about climate change and want to make a difference. Sometimes it can be overwhelming when you feel so passionate about such a deeply troubling problem like our warming planet and it seems like no one around you is concerned, especially in a state that sometimes feels hostile to your efforts. Finding CCL is a breath of fresh air to our volunteers—yes, there is a local organization that cares about the issues that you care about!

CCL focuses on optimism and relationship building. Our volunteers are provided with a lot of training to learn how to find common ground with people who don't always agree with us. For example, when we talk to conservatives about

renewable energy projects, we emphasize job creation, energy investments and opportunities to outcompete China with investments in Texas. Not only that, we also provide practical actions our members can take to fight climate pollution and prepare ourselves for the effects of our changing climate. We build relationships with our elected leaders and support efforts at the state and national level to get good climate policies passed. When a chance comes along to make our world more livable, we jump on it and do what we can to help."
Cheryl Clark, volunteer with the DFW chapter of the Citizens' Climate Lobby

TWO
Tipsy Tightrope: Losing My Marbles

In the midst of doing a million things and nothing at all, my phone chimes. I drop everything and nothing to check it. Of course. My phone is my mistress. I smile when I read, "Hello Darlink." It's a text from my friend, Joie. "We haven't spoken since the Supreme Court's decision limiting the EPA's regulation of carbon emissions from power plants. I just hope you're breathing and functional. If you're on the floor hopefully you're relaxing. More art."

Smile fades. "I am completely undone," I text back. "But I expected nothing less," I add with the practiced stoicism that keeps me upright these days. It's the only thing keeping me from pulling out my own fingernails. I do appreciate her checking in with me. I'd be worried, too, if I were my friend.

As the Supreme Court assails all that I hold dear, I feel I'm walking a tipsy tightrope, the grip on my balancing pole white-knuckled and my legs shaking like a 20-cup-a-day coffee addict. On one side lies total collapse, the fiery pit where I'd land if I allowed myself to feel the full horror of what we've done to our lovely planet. Not "weather," not "natural variations," but destruction wrought by *homo sapiens sapiens*—the great irony being that "sapiens" means "wise." We've named ourselves "wise," not once, but twice. Can't make this stuff up. Too bad we let money make us stupid.

My rants against God on the topic notwithstanding, we not-so-sapiens have known what we're doing for decades now. Charles David Keeling's measurements of atmospheric levels of CO_2 at Mauna Loa in the late '50s proved that burning fossil fuels was increasing the levels of CO_2 in the atmosphere at an alarming rate. Rachel Carson informed us about the deadly impacts of putting poisons like DDT into the environment in *Silent Spring* in 1962. Heck, George Perkins Marsh wrote about the deleterious impacts humans make on our environment in *Man and Nature*, published in 1864. One hundred and sixty years ago...

The ravages committed by man subvert the relations and destroy the balance which nature had established between her organized and her inorganic creations . . .
The earth is fast becoming an unfit home for its noblest inhabitant, and another era of equal human crime and human improvidence. . . would reduce it to such a condition of impoverished productiveness, of shattered surface, of climatic excess, as to threaten the depravation, barbarism, and perhaps even extinction of the species.

George Perkins Marsh

Tell me again how "we just didn't know."

Legitimate climate scientists may disagree about the timing of when our climate will be too hot to support our kind—but not about whether it's going to happen. Not, that is, if we don't make some major changes, and fast, preferably yesterday. Even better, twenty or thirty years ago. Will we reach that point in the lifetimes of my children or grandchildren? Let's just cross our fingers and hope not! Since the EPA can't tell companies to stop burning coal,[30] we just hope for the best.

But on the other side of my tightrope lies the zoned-out numbness that keeps me from feeling anything at all—not even the joy of my child's hug or the delight of a hummingbird sipping nectar from a nearby flower. Avoid the horror, the crippling grief, the paralyzing depression inspired by our seeming refusal to pick up the tools to save ourselves. Turn on *Ted Lasso* and escape the world. Pass the Ambien, please.

But none of that soothes me when I wake up screaming in the night.

Stemming the rush of the roaring river of my terror-filled thoughts is not easy, not with this overactive Gemini mind of mine. It does not like to take time off—its vast stash of unused PTO[31] languishes in a dark closet full

---

[30] Or, thanks to cases decided in 2024, limit smokestack pollution that blows across state borders. Or protect millions of acres of wetlands from pollution. And the reversal of the "Chevron doctrine," which gave certain deference to agency-enacted rules, means the Supreme Court is just getting started. Look for a new slew of cases challenging environmental protections on their upcoming dockets. WHILE I SCREAM.

[31] Paid Time Off, which is definitely a joke because I don't get paid for overthinking. I'd be a billionaire if I did.

of spiderwebs. Which could be any closet in my spider-friendly (*aka too-lazy-to-clean*) home.

So, I do what I can. I've upped my meditation practice to a regularity I normally reserve for opening the refrigerator to get cheese. I slog myself to the gym with far less bribery than I used to require.[32] I force myself onto the stand-up comedy stage when I really need to rant (without further alarming my family). And I make a point to read some "feel good" news from *YES! Magazine* or check in with the cute bears on the National Park Service socials.

But does any of that keep permafrost from melting and thereby releasing billions of tons of greenhouse gases? Does it stop companies firing up their coal-fired plants? Does it keep any of us from hopping into our gas-guzzling, emission-spewing vehicles?

No. It does not.

It does, however, get me through another day without mad-shrieking at random people on the street. I used to give a wide berth to the self-appointed public pontificators. Now, I feel a sense of camaraderie. Here are my people.

If only there was a government entity specifically charged with protecting our natural environment. Oh, wait, there is! Congress established the Environmental Protection Agency some 52 years ago, on the insistence of that *Republican* U.S. President Nixon, no less. But apparently, giving an agency the mandate to "establish

---

[32] Who thought it was a good idea to put a Wetzel's Pretzels just outside my gym door, anyway?

and enforce environmental protection standards consistent with national environmental goals" wasn't clear enough to empower it to tackle crises of such epic proportions.

Too bad. Climate change is already wreaking havoc across the United States—even more so around the world. Heat waves are more intense now— the summer of 2024, as I'm writing, being particularly brutal—July 21, 2024 was the hottest day ever recorded . . . until it was edged out by July 22, 2024.[33] Can't make this shit up.

And these aren't just flukes. As of this writing, each of the prior thirteen months was the hottest on record, the average heat wave season in the United States being about 49 days longer now than in the 1960s.[34] Rising seas are already resulting in more frequent flooding, and submersion of land. The Southeast, from North Carolina to Florida, lost about 20 square miles of land between 1996 and 2011.

Ticks and mosquitos love the warmer weather and shorter seasons, so they are becoming more numerous, bringing more Lyme disease and West Nile virus, spreading into areas not previously hospitable to their kind. More heat means more "killing-degree days" for crops when temperatures are too scorching for their survival. It's also bad for their pollinators who depend on temperature and precipitation cues to know when to pollinate—some pollinators are already out of sync with

---

[33] Seriously wondering how many times I'll need to update this footnote before publication, seeing as how the heat waves of the summer of 2024 continue, as of this writing.

[34] I have no idea what I was going to say here. That ever happen to you?

periods of plant blooming. More devastating firestorms rage across the West in the U.S., while the wildfire season extends longer and longer.

But don't hold your breath waiting for the EPA to step in and fix any of that. The U.S. Supreme Court has them bound and gagged.

Not that I'm bitter.

Actually, scratch that. I am so effing bitter that my saliva tastes of dandelion greens soaked in lemon vinegar. Even my pacifist self has rolled up her sleeves and wants to slug somebody. Perhaps somebody who wears a black robe to work.

Okay, okay, I take that back. I'm not one to incite violence. And I do pray to learn how to communicate with people whose seeming heartlessness infuriates me. Because if we don't learn how to talk with each other, with reason and consideration, about the crippling issues of the day?

Alarm bell rings

It's time for my art therapy. I'll let you decide how to answer that question.

"I think about climate basically all the time. And one thing I wonder is why the Earth getting irreversibly hotter just doesn't seem all that scary or even urgent to most people. To me it's truly horrific. I wonder if my brain just works differently or something."
Climate scientist Peter Kalmus

BUREAU OF LINGUISTICAL REALITY
"A dictionary for the future present"

Ever feel like you lack the words to express the frustration, the angst, the worry, the fear, the fury, the whole slew of emotions that environmental crises inspire? Me, too.

Maybe it's time to coin some new ones.

That's where the **Bureau of Linguistical Reality** comes in. Founded by Heidi Quante and Alicia Escott, the Bureau invites us all to create new words that better reflect our experiences in our rapidly changing world, thanks to climate change and a host of other crises. The goal is to develop the language to facilitate the kinds of conversations that we need to be having right now.

What kind of new words? Here are few examples:

"Ennuipocalypse

Definition: While media often depicts the apocalypse as a sudden and dramatic event, the Ennuipocalypse, or Slowpocalypse (slang) offers the concept of a doomsday that occurs at an excruciatingly slow day-to-day time scale.

> **"Shadowtime**
>
> **Definition:** A parallel timescale that follows one around throughout day-to-day experience of regular time. Shadowtime manifests as a feeling of living in two distinctly different temporal scales simultaneously, or acute consciousness of the possibility that the near future will be drastically different than the present."
>
> Find more information about these new words, as well as many more at the Bureau's website. Or submit your own!
>
> https://bureauoflinguisticalreality.com/

THREE
The Band Played On

"A beloved juvenile red wolf named Muppet has been killed by a vehicle strike . . Muppet, who was apparently struck and killed on April 15, was the fourth red wolf road mortality in the past 10 months. Muppet's father was also killed by a vehicle strike six months earlier . . . Fewer than 20 red wolves remain in the wild, making them the most endangered wolves on the planet."

Sad stories like Muppet's flood my inbox, making me certain I lack the emotional fortitude to stay well-informed anymore. These emails now pile up because I'm on a News Diet, one which I pray will be more successful than any of my food restriction efforts.

Just yesterday, another awful email slipped through my definitely-not-ironclad blockade, informing me that Earth's operating systems are failing—on six out of nine fronts, in fact—and two of the other three are hanging by a gossamer thread. Which sounds as encouraging as your doctor saying your blood pressure is too high and your cholesterol is astronomical, but your lungs may be okay for another six months if you quit smoking. Except this scenario is more akin to demanding 8 billion people quit smoking[35], while the governments of the world pass out subsidies to cigarette manufacturers—but what's lighting up the Internet is the Tide pod challenge.

"We are in very bad shape... We show in this analysis that the planet is losing resilience and the patient is sick."

Johan Rockstrom, director of the Potsdam Institute for Climate Impact Research in Germany.

The more I read about the collapses of our Earthly life support systems, the more I think about this phrase: "and the band played on." I picture the eight-member band[36] aboard the *Titanic* playing ragtime and hymns[37] to soothe the freaked-out passengers as the ship groaned into the ice-cold, inky depths. Hearts pounding, perhaps struggling to stay in their seats as their chairs slid when the pitch of the ship tilted steeply, still doing what they loved best. I consider them lucky, as compared to the other ill-fated passengers, anyway. In their final

---

[35] As in STOP BURNING FOSSIL FUELS.
[36] For posterity: the members of the band were violinists, Wallace Hartley, George Alexandre Krins and John Law Hume; violist and bassist John Frederick Preston Clarke; cellists John Wesley Woodward and Roger Marie Bricoux; and pianists Percy Cornelius Taylor and Theodore Ronald Brailey. None survived the sinking of the *Titanic*.
[37] Many accounts claim the band played "Nearer My God to Thee" at the end, but this hymn wasn't in the songbook reported to be aboard the *Titanic*. The 22-year-old junior wireless operator, Harold Bride, identified the song as "Autumn," which is the tune for a variety of hymns popular at the time. In case you were wondering.

moments, they had something to give, something to do to offer solace, something to do that gave them joy.

I honor their devotion as I consider my own role in the unfolding catastrophe of our time. I think of them as I crank up the tunes to drown out my own sense of impending doom. I think of them as I do what I know how to do and what gives me joy, be it ever so small or ineffective considering the looming eco-apocalypse.

I want to delve into the findings of that disturbing "Earth Beyond 6 of 9 Boundaries" study—which I'll call the "Earth Out of Bounds report" in the remainder of this book—and get a better understanding of what these assessments mean. If the "doctors" have issued a dire prognosis for our Earthly life support systems, shouldn't we at least read their reports? My children's futures depend on the viability of these life support systems, after all.

I've managed to summon the courage and the necessary fortifications to delve into the report, and to devote an entire chapter to each topic. Seems the least I can do, given the gravity of the topics. Note that I'm not claiming to represent the exact findings of this scientific report—some of it is over my head, I confess. I'm sharing what I got out of reading about each category of concern, what fears and possible responses each prompted.

Still, I am worried about putting any additional strain on my own emotional support systems. I'm already at DEFCON 2, with little provocation required for me to go nuclear.

So, I've taken a cue from the musical therapists of the *Titanic* for the chapters in which we consider the "Earth

Out of Bounds" report. I've created a playlist of suggested musical selections to play as we go,[38] to ease our collective distress, as we consider all this awful news.

Taking another cue from my own eco-grief management plan, I'm also dispersing the "Playlist chapters" among some other, more frivolous stories. Let's give ourselves some time and space to process the worrying conclusions of the "Earth Out of Bounds" report. To grieve, to curse, to rage before plowing through the next installment of furious facts. There's always time for cursing.

Facing the truth about the state of the natural world that supports all our lives is not for the faint of heart. I know. I've spent many a dark day, cowering and crying. Seeing my children begin their young adulting lives, knowing the horrific crises they are inheriting, convinces me to step up my efforts to do what I can.

My book *Love Earth Now* focused on our discovering and heeding our individual calls to action—along with essential self-care for body, mind and spirit. All are vital if we are to have the bandwidth to stay engaged.

This time around, however, I'm broadening the perspective into the sphere of our cul-de-sacs, communities, clubs and confederations. Much as the Powers That Be may want to keep us fragmented and isolated, we can do so much more working together. From our places of despair, we can take heart knowing how many community-based, national, and international organizations are already taking concerted

---

[38] Listen on Spotify https://tinyurl.com/yb9dd4hm OR Tidal: https://tinyurl.com/ycx4kvff. Or search for "And the Band Played on For Madness Readers"

action. They offer us not only hope, but camaraderie for the long road ahead of us.

I'll still do some of my trauma processing alone by mad-shrieking in my closet, but this introvert has discovered a deep satisfaction in collaborating, pooling resources and taking turns. When I need a break, I'm grateful to know the work continues. Grassroots multitasking to the sanity rescue.

Deep breaths . . .

"If we wait for governments, it'll be too late; if we act as individuals, it'll be too little; but if we act as communities, it might just be enough, just in time."
Rob Hopkins, cofounder of the Transition Town Movement.

TRANSITION TOWNS

"Community-led Transition groups are working for a low-carbon, socially just future with resilient communities, more active participation in society, and caring culture focused on supporting each other. . . . There is an increasing recognition that top-down approaches are not sufficient alone to affect change and need to be combined with community-level responses."

https://transitionnetwork.org/
If you're looking for your "eco-village," folks in your local

area who are building a more resilient place to live, check out the Transition Town Movement. Launched in 2006 in Totnes, England, community members came together to re-imagine what it means to live in a sustainable society. Not waiting for their government, big corporations, or anyone else to implement it for them, they selected their own projects, designed to support their local community in adapting to a rapidly changing world.

The idea spread to other local communities, and now there are Transition Town chapters in towns, villages, cities, and schools across 48 countries. Is there one near you?

Each Transition group sets its own priorities, agendas and actions, guided by principles and resources from the organization. They may opt to focus on local food production, energy dependence, pursuing zero waste goals, community education, developing a local currency, and more.

How do you start a Transition Town? Check out the TED Talk of Lynn Hartle, a member of the Transition Town in Greater Media, Pennsylvania.

https://www.ted.com/talks/lynn_hartle_transition_towns_jan_2022?

"In Transition Town Greater Media (TTGM), we launched a FreeStore which was created to keep household items out of the landfill, create a circular economy, and help people understand how a "gift economy" can contribute to reducing our waste stream and improve community resilience. We also helped foster the Green Wagon project, which is like a little lending library, but for native plants. It provides a place where neighbors can get free plants for their yards, participate directly in improving the neighborhood biodiversity, and learn about the benefits of planting natives. Both initiatives promote opportunities for socializing, connecting and learning while helping our community thrive."

Skip Shuda, TTGM President, Media PA

Learn more about the Transition Town in Greater Media, PA:

https://ttgmPA.org

FOUR
Playlist Track 1: Biodiversity

♫ **Recommended Musical Pairing** ♫
"What's it All Mean?"
The Philharmonik
Winner of NPR's 2024 Tiny Desk Contest.

The "Earth Out of Bounds" report outlines a "planetary boundaries framework" of nine processes deemed critical to maintain the support systems necessary for Life as We Know It. As essential as our human respiratory, cardiovascular, digestive and stressed-out nervous systems, we need these nine Earthly processes well-oiled and humming to ensure an environment that remains hospitable to our kind.

One of the already-exceeded measures, "biosphere integrity," assesses the health of the biodiversity on Earth. What's biodiversity? Does it have anything to do with DEI, that diversity, equity and inclusion acronym that's inspired so much acrimony of late? You can

PLAYLIST TRACK 1: BIODIVERSITY

breathe a sigh of relief because the answer is no. We already have enough controversial topics to tackle.

According to *National Geographic,* "Biodiversity refers to the variety of living species on Earth, including plants, animals, bacteria, and fungi." In other words, it means we aren't the only species on Planet Earth that matters. I do hope that doesn't come as a shock to anyone.

As you might surmise from the sad demise of Muppet, that endangered red wolf, we're not the best neighbors. One in eight species on Earth —that's about a million species, I repeat ONE MILLION—faces possible extinction.[39] We've already caused the extinction of some 83% of all wild mammals and half of all plants—most all of which lived and thrived here on Earth eons before we humans arrived. Which makes us the worst possible roommates. Can we please learn to share the planet?

Oh, but I hear someone insisting that the extinction of species is a "natural occurrence." Indeed, it is, as evidenced by the lack of T-Rexes and trilobites in our modern world. Since life began some 4-plus billion years ago, Earth's living inhabitants have taken a variety of forms, their populations growing and vanishing, like the ebb and flow of the tides. Scads[40] of species have gone extinct because they couldn't adapt to the variations in the climate and atmosphere, and scads more because of cataclysmic episodes that we humans weren't yet around to cause. Even when some 96% of all species died off 250

---
[39] Estimate based on analysis of some 15,000 studies conducted within the last 50 years. In other words, NOT A HOAX.
[40] A precise scientific term meaning I have no idea how many.

45

million years ago,[41] the life forms that survived learned a thing or two, and Life went on.

If this all sounds reassuring—yes, yes, extinctions are natural—it's because we've forgotten a few critical distinctions: none of those gone-extinct species were the cause of their own demise . . . while simultaneously denying they were doing it . . . and refusing to do anything to adapt to the very changes they were causing.

What makes our situation even more alarming is how much we've accelerated the natural rate of extinction with our nature-destroying habits. Scientists say, over the past ten million years, species have gone extinct at a rate of one to five per year—but extinctions are now occurring at a rate tens or hundreds times higher, depending on who you ask.

I don't know how you feel about life getting faster and faster, but I hear people complaining about it every day—and that's not just me when I'm running perennially late. Shouldn't speeding up the rate of *deaths* be even less popular? Only if we were paying attention to it. Or cared about any species other than the faces in our mirrors.

But, so what, right? So long as we're talking about red foxes or wild yams disappearing, who cares, so long as everyone has Netflix and an iPhone.

No matter how hard we, in the U.S. anyway, pretend that we're just individuals pulling ourselves up by our bootstraps, our entire existence depends on a multitude of other species. How many species do we require to feed

---

[41] Increased CO2 levels due mostly to volcanic activity likely caused the mass extinctions. So why, I ask, would we do this to ourselves, increasing CO2 levels possibly to the point of our own extinction?

us, pollinate our food's food, digest our food, clothe us, recycle our wastes, turn over our soils, purify our air and water, provide us with building materials and precious minerals for our iPhones, heal us with life-saving medicines, and let us not forget, *entertain* us?[42] I'm not sure, but let's go with SCADS.

Time to crank up the tunes.

So what song am I singing? What can I do to help preserve what biodiversity remains?

Come with me to my front yard overgrown with native plants. Bees, butterflies, and hummingbirds flock to the native sages, buckwheat and laurel sumac. Fluorescent green beetles, pillbugs, and a myriad of microbes call my compost bin home. All buzzing, munching and (from what I can tell) thriving with nary a drop of pesticides or herbicides applied.[43]

As much as I relish in the biodiversity I see in our wild yard, I know it's but a drop in the healthy ecosystem bucket of what's needed to ensure that our local species thrive. Come with me now to the Wallis Annenberg Wildlife Crossing, currently being constructed over all ten lanes of the US 101 freeway in Agoura Hills, California. When completed, it will be but one of a thousand wildlife crossings across the U.S.—and the largest. The top of the crossing will feature nearly an acre of habitat full of native plants, and it will offer shelter,

---

[42] I shudder to think where I'd be without clumsy cat reels and stupid pet trick videos. Life-saving therapy these days.
[43] I am, of course, omitting any mention of the other plants I've killed off, but I'm content to share this land with the species who are content with my laissez-faire style of gardening.

food and water. The massive freeway below will be out of sight and earshot of the critters crossing it. That's a lot more habitat than found in my front yard, for sure.

This crossing has been years in the making and is sorely needed to protect local biodiversity. Some 300,000 to 400,000 vehicles traverse this stretch of the freeway as it passes along the north side of the Santa Monica Mountains in Agoura Hills, just northwest of Los Angeles. The Sana Monicas stretch east-west some 40 miles from the Hollywood Hills to Point Mugu in Ventura, California. Completely surrounded by freeways and bisected by freeways (and the Pacific Ocean at one end), the Santa Monicas are a biodiversity island. Critters living in the area have no safe access to the largely undeveloped Simi Hills and the vast Santa Susana range to the north and beyond. Inbreeding is inevitable when geographic range is so limited, and that's bad for biodiversity.[44] It's of particular concern in this area which has been recognized as one of only 36 biodiversity hotspots worldwide.[45]

A cougar dubbed P-22 famously crossed both the vast I-405 and the 101 freeways to make his home in the Griffith Park area of the mountains.[46] Griffith Park encompasses some nine square miles in the heart of the

---

[44] Ask the royal families how well inbreeding has worked out for them.
[45] And let's give thanks to Fran Pavley, an environmental leader and former California legislator, that it hasn't already been plowed under for more mini malls. In the early 1980s, she led the charge to preserve this land which remains the only undeveloped acreage of this size along the 101 freeway corridor.
[46] Many other mountain lions have died trying, making P-22 all the more extraordinary. A University of California Davis study found that, on average, one mountain lion was killed on a road in California every week between 2015 and 2022. Somebody hand me a box of tissues.

city, making it a natural treasure for Angelenos—but a habitat 31 times too small for an average male mountain lion.

For some ten years, P-22's tenacity and piercing gaze captured the hearts of Angelenos who mostly forgave his occasional forays into their backyards. Weakened by injuries from a car collision and infections, P-22 was finally euthanized in December 2022. Los Angeles grieved.

A moment of silence for our courageous cougar neighbor. P-22's death was a catalyst for action, spurring the long-proposed wildlife crossing to become a reality.

And now a word of gratitude for the 5,000 individuals, foundations, agencies, and businesses around the world whose financial donations, along with countless hours donated by volunteers, are making this wildlife crossing a reality. This project demonstrates the power of many coming together to work for a common good.

And if the plight of some urban mountain lions in a city where you do not live fails to stir you, consider the million car-and-wildlife collisions every year in the United States, killing 200 people and causing $8 billion in losses.

May we have more wildlife crossings, please?

#ShareThePlanet

You can view the construction progress on the webcam here. https://101wildlifecrossing.org/

"We need to move beyond mere conservation, toward a kind of environmental rejuvenation. Wildlife crossings

are powerfully effective at doing just that — restoring ecosystems that have been fractured and disrupted. It's a way of saying, there are solutions to our deepest ecological challenges, and this is the kind of fresh new thinking that will get us there."

Wallis Annenberg, Chairman of the Board, President and CEO of the Annenberg Foundation, a generous donor to the Wallis Annenberg Wildlife Crossing project.

THE SOAP BOX:
RAT POISON KILLS MORE THAN RATS

Some 95% of mountain lions tested in California have ingested rat poison. If it doesn't kill them, it weakens them, so they succumb to other causes. The National Park Service treated P-22 for rat poison ingestion in 2014 (he'd been collared after his discovery in 2013), likely extending his life some ten years.

Rat poison doesn't just kill rats. Predators consume the poisoned rats, thereby threatening owls, eagles, falcons, snakes, weasels, bobcats, cougars, foxes, coyotes, badgers and our pet cats and dogs. People, especially young children, are at risk too, if they get bitten by poisoned rat. Rat poisons prevent blood from clotting, so critters die of internal bleeding, a truly horrendous way to go.

Check out these resources to learn about safer means of rodent control:

The Hungry Owl: www.hungryowls.org/
RATS: raptorsarethesolution.org/
Poison Free Malibu: poisonfreemalibu.org/

U.S. Department of Transportation Wildlife Crossing Program

Wish there was a wildlife crossing where you drive? The Infrastructure Investment and Jobs Act of 2021 created and funded the Wildlife Crossing Program, a pilot program offering discretionary grants to reduce wildlife and vehicle collisions while improving habitat for both terrestrial and aquatic species.

For details, see:
https://highways.dot.gov/federal-lands/wildlife-crossings/pilot-program

WANT TO DO MORE TO PROTECT BIODIVERSITY?

Look for a **Ducks Unlimited** chapter in your community, your university, your high school.

If you're scratching your head, wondering, but isn't that a "hunting organization"? How does hunting help biodiversity?

Hear me out. Indeed, Ducks Unlimited ("DU) was founded by hunter-conservationists in 1937 to conserve wetlands and associated habitats for waterfowl and other wildlife. To date, DU has conserved over 18 million acres of habitat across North America.

No small feat when you consider how critical wetlands are for the one third of all birds who rely on wetlands for nesting, feeding, and/or migration. ONE THIRD.

Wetlands provide countless environmental benefits, not only for waterfowl, but for fish, amphibians, shellfish and insects. Many species rely on wetlands for food, water and shelter. They store carbon within the plant communities they support, filter water for purification, and help protect against flooding and shoreline erosion.

Who better to protect the wetlands and related habitats than those who know them best, spending hours wading and observing in the marshes, fens and bogs? Those whose beloved pastime would vanish if not for healthy waterfowl ecosystems?

I queried Dr. Mike Brasher, DU's senior waterfowl scientist, as to why non-hunters should support the organization.

"Hunters have been at the forefront of conservation for decades, and not just because of the billions of dollars contributed through annual license sales and excise taxes on the equipment and firearms used by hunters. But also because of the connection they have to the resource and the efforts they put into protecting and managing their own lands to provide habitat for waterfowl and other wildlife.

Groups like DU aren't just for hunters. Many of our members don't hunt, or maybe they once did but have since transitioned to other forms of recreation. Ducks Unlimited is better viewed as a family of passionate conservationists that aspire to contribute to something bigger than ourselves. People can get involved in DU in many different ways, but the easiest is to check out one of our local fund-raising events at www.ducks.org/events *and see if DU is right for you."*

If you're anti-hunting and just can't support a group like DU, find a different outlet. But I challenge us all to find ways to raise a bigger tent. To embrace a common bond without insisting that everyone go about creating a more sustainable human presence in the same way. If we are to survive as a species, we must find ways to bridge our gaps, to work together to find solutions that work for not just us and our compadres, but for Life on Earth. Thank you for coming to my TED Talk.

Read more about the conservation efforts of Ducks Unlimited here.
https://www.ducks.org/conservation/conserving-wetlands-waterfowl

FIVE

Make It Hard: Lessons In Letting Go

Today, I had a showdown with my garden hose. I'm not proud of it, but I'm in a confessional mood. I'd like to say I won, but that would be like saying I beat myself in a fist fight. With myself.

Because didn't I create the whole altercation? I mean, sure, the blasted thing was snarled around its rusting reel and refused to unfurl just six more inches—JUST SIX MORE INCHES! How hard is that? That's all I needed to reach the lovely native sage I recently planted, which was Totally a Good Thing. By this planting, I introduced another native plant to delight the local fauna, the hummingbirds and the butterflies. I'm so overcome with my own righteousness, I gotta wear shades to shield my eyes from the sanctimonious glow. Why can't the bleeping garden hose acknowledge my virtuousness and help me out just a freaking little?

Sure, it sounds ridiculous to you, because it is, but it all makes sense if you are me. And I don't blame you for being glad you're not.

Because I like to make things hard. The simpler the task, the more I take perverse pleasure in making it more difficult, thereby giving myself a basis for indulging in my favorite pastime: complaining.

Did I take five minutes to untangle the garden hose before using it? I did not. Because that would deprive me of the aforesaid perverse pleasure of grumbling about my First World problems. The ones I create. And endeavor to make harder. Which makes retelling the story of the whole calamity all the more dramatic. Over a nice frosty mug.

Funny thing is that you'll rarely hear me complain about the Big Stuff.[47] Not enough money to pay this month's bills? Radio silence. I got diagnosed with cancer? I'll regale you with the horrors of my hangnail. Terrified we've messed up the planet so badly that my children will never know the natural beauty and stable climate I've taken for granted? Shove that down as far as it goes.

Tangled garden hose? Let me sit right down and compose a long, bloated essay about that. I suppose it's easier to deal with the small stuff, to know where to draw a line in the sand, delineate what's good and what's bad, to find the holy ground. The garden hose is either tangled or it's not. The sewing machine is either making a fine seam or the thread broke mid-stitch, without any provocation whatsoever, while I obliviously motored on. Who does that? Grr. I force myself to plow through a year's worth of procrastination just to take up the mending at all. When I do, I'd appreciate some

---

[47] This essay was obviously written well before I started this book.

acknowledgement, some cooperation, from the machine that would otherwise sit collecting too much dust.

But facing the awful truth about the many ways my lifestyle contributes to the death of our life-sustaining environment? That every time I crank up the air-conditioning in the ever-hotter summer weather, another chunk of ice calves from Antarctica? That another mountain lion died trying to cross the freeway I've driven so often?

The trouble with these big-issue worries is that they aren't solved as easily as untangling a garden hose. Even if I give up driving and flying and air-conditioning, millions of other people would still be cranking up their ACs, racing down the freeways in Hummers, powering up their ginormous yachts and private jets. Would my monumental (to me) sacrifice make any difference at all in the grand scheme of things? Is there ANYthing I could do to spare mountain lions from extinction or Pacific islanders from losing their homes to rising seas?

Much as I preach "small things matter," I confess I have my doubts, too. We, at least those of us in the developed world, have all played a part in wreaking the destruction of our life support systems. Extracting whatever we want, by whatever means, and wantonly discarding our trash is built into every aspect of our cultural, social, governmental and financial systems. Me sweltering in my home isn't going to change all of that, no matter how much I wish it would. And right now, I have no idea what I can do that will.

Which sends me running back to the garden hose situation. It's so much more manageable, within my control, even if, in my snit, I pretend otherwise. A simple apology would've sufficed, but it's got nothing to say for itself; it just sits there flaunting its snarl. I give it a kick, and no one's feelings, except for those in my big toe, are hurt by that. Garden hoses hold no grudges.

Slumped now against the garden bench, I see the desiccated remains of a once thriving plant, waving in the breeze, as if waving a white flag, pleading for sustenance on this scorching hot afternoon. Eventually, I'll take pity, summon my only smidgeon of patience to untangle the hose and give us all a break from my pointless berating of inanimate objects.

Just as soon as I'm finished chewing out the ice maker. Must it jam up on the hottest day of the year? I grab a sharp knife to stab at the clogged ice. This is going to be fun.

This is your century.
Take it and run as if your life depends on it."

Paul Hawken, environmentalist, author, activist

SIX
Playlist Track 2: Plastics

♫ **Recommended Musical Pairing** ♫
"Where Do the Children Play?"
Cat Stevens

"Researchers discovered that median levels of some microplastics were over 10 times higher in baby feces than in that of adults. Is it surprising given their toys, their clothes, their cribs, their playgrounds, breast milk, can all contain, leach or shed microplastics? Babies and infants are inhaling and ingesting microplastic particles at nearly every stage of modern life."
Babies v. Plastics report, Earthday.org

Another already-exceeded measure in the "Earth Out of Bounds" report concerns "novel entities," an innocuous-sounding term for the truly noxious stuff that humans have created, like nuclear waste, synthetic chemicals and plastics. How badly have we bollixed the planet by spewing all these poisons? The *tl;dr*[48] summary is this: there's no safe level for any of this crud, so we should just stop making it.

Good effing luck with that. Plastic is so pervasive in our environment that some scientists have dubbed our current epoch "the plasticene."[49] Microplastics have been found in the depths of the oceans, atop the highest peaks, in our food and drinking water, and in our living bodies. It's no wonder, I suppose, given how much plastic we generate, use and trash daily. How many goods, services, activities or places can we name that are untouched by plastic in some form? Take as long as you need.[50]

I'm not even going to pretend any significant amount of all that plastic gets recycled. The United Nations Environment Programme estimates that we humans manufacture some 430 million tons of plastic each year—weighing more than all eight billion people on Earth. And a whopping (*sarcasm*) 9% of all the plastic ever produced has been recycled. A *truly* whopping 72% ends up in landfills or elsewhere in our environment. A truckload of plastic trash lands in our oceans every Hoover dam minute. And if you haven't seen the heartbreaking videos of the albatross

---

[48] "too long, didn't read"
[49] Not to be confused with the Pleistocene, the epoch of the most recent Ice Ages, from 2.5 million years ago to 11,700 years ago. The current epoch is officially called the Holocene. But just to confuse things further, some scholars have proposed a new term, the Anthropocene ("anthro" coming from Greek for "human"), to reflect how radically humans have altered the Earth. While I highly recommend John Green's book, *The Anthropocene Reviewed: Essays on a Human-Centered Plane*t, it's not an official designation.
[50] Read *A Year of No Garbage* for a glimpse of just how near-impossible living without single-use plastic is. And don't get me started on how we don't even talk about the disposal of so-called "durable" plastics, the ones that get used more than once, but what then? Do those just magically vanish?

parents on Midway Atoll feeding plastic bits to their starving young—because that's what they scoop up in their bills from the ocean now—I urge you not to look them up unless you have a stronger constitution than mine. Their plaintive peeps haunt me in the night.

I have yet to hear a plastic manufacturer say sorry, let alone offer to clean up their messes. Have you? We teach our toddlers better manners than that.

If I chuck a plastic bottle out of my car window (à la mode when I was a kid in the early 1960s), I've committed a littering misdemeanor, subject to a fine. But we can't discourage the mass production of toxic plastic, which already trashes every corner of the planet (and our bodies), because it's bad for the bottom lines of the petrochemical companies?

Plastics aren't just a problem for albatrosses or those sad sea turtles caught in plastic six-pack rings. Plastics contain any number of some sixteen thousand man-made chemicals, some 25% thought to pose high hazards for people and the environment. A lot of those chemicals are either a "persistent organic pollutant" or a heavy metal, like highly toxic lead, mercury, and chromium.[51]

What's a "persistent organic pollutant" (and its cute-sounding acronym, "POP")? Some POPs, like DDT and PCBs, we outlawed long ago because we knew they were awful. But new ones keep "popping" up (pun intended), in pesticides, insecticides, solvents, pharmaceuticals and industrial chemicals.

---

[51] Not, in other words, heavy metal like Ozzy Ozbourne no matter how much his "Crazy Train" song feels apt right now.

They are like the worst-ever house guests: not only do they never leave, but they also destroy the place after they move in. They travel by air and water, too, so they are roommates that leave a trail of destruction far beyond your apartment. What's so bad about them? POPs get stored in our fatty tissues and proceed to inhibit normal cell functions, causing problems for everything from normal growth and reproduction to causing cancer. Fun guests.

But we'd never let such noxious stuff into our homes, would we? So long as we keep our plastic water bottles out of the sun and our Tupperware out of the microwave, it's all good, right? Those are the only warnings about plastic that I'd ever heard before—keep them out of sunlight and heat.

Incorrect. Plastics "shed" worse than a Persian cat in a heat wave. Scientists have found some 400 different plastic chemicals in a bottle of water, just twenty-four hours after it was filled. How long does water sit in a plastic bottle before we purchase it? They found some 3500 plastic chemicals after cleaning said plastic bottle in a dishwasher and refilling it. Suddenly, I'm reconsidering those so-convenient microwavable meals that come in a plastic tray... which we heat in the microwave ... Why don't these things come with warning labels??

What's worse is that recycling is a cruel joke. At best, only plastic containers marked with the "chasing triangle" numbers of 1 and 2 get recycled—the vast majority of plastic produced today goes straight to the leaking, methane-emitting landfills. And the plastic that does get recycled? That's only an option once, maybe twice, before degrading into unusable waste and becoming "forever trash."

Since plastic never breaks down into anything like compost, what happens to discarded plastic? It just breaks down into smaller and smaller bits. Microplastics, smaller than a grain of sand, and nanoplastics, invisible to the naked eye, accumulate in our bodies, in our organs, in human placenta and breast milk. Even in our clogged arteries. Plastics in women's bloodstreams may be linked to increased risk of first trimester miscarriages.[52] And, yes, gents, microplastics have been found in testicles and penises, possibly leading to decreased male fertility. Got your attention now?

If you'd asked me whether humans could cover the Earth and every living being with any one thing, I would've said it's impossible. Yet here we are, "succeeding" beyond my wildest imagination. Why, oh why couldn't it have been critical thinking that we exported to every corner of the planet?

But back to babies, the smallest and most vulnerable of us all. When I think of all the plastics in baby blankets, clothing, disposable diapers, toys, pacifiers, bottles, even the carpets they crawl on (and dust bunnies, too), I am not surprised that the level of plastics in their feces were ten times that of an adult's. TEN TIMES! I'm horrified,

---

[52] Which is even more horrifying when you consider how many women, in these "post Roe v. Wade times," are being charged with crimes after a miscarriage these days. See the case of Brittany Watts, a Black woman in Ohio, who was charged with "abuse of a corpse" after she miscarried a non-viable pregnancy at home. Doctors had refused to treat her and sent her home. The grand jury declined to pursue charges, but what a nightmare for someone already traumatized by enduring this painful process without medical support. I digress. Again.

saddened, disgusted, and infuriated, but no more surprised than finding a tortilla chip in my hand when I'm on a diet.

The countless hours I spent baby-proofing our home now seem a joke. All those soft blankies, onesies and plushies were shedding chemical-laden nanoplastics like a drug dealer when the cops show up.

And let's hope their toys aren't made from recycled e-waste, because another study found that they can expose children to highly toxic chemicals. I used to consider plastic toys safer than the sharp-edged metal toys I inherited from my mom's youth. What's a sharp corner compared to a constant diet of toxic chemicals leaching from everything the kids touch?

No wonder the sea turtles don't stand a chance; we aren't even protecting our most precious ones. Of course, if we really cared about the littles, we'd definitely do something about gun control, right?[53] Sorry, wrong rant.

But worst of all, and yes, it can possibly be even worse: the fossil fuel and other petrochemical companies that make this shit knew the truth: recycling was never a realistic solution. A recent study found that plastics recycling was a scam that the petrochemical companies foisted on consumers, on cities, and us eager do-gooders who carefully rinse and sort our yogurt cups. They always knew it was a feel-good panacea, never meant to be technically or economically viable. *Scads of expletives deleted.*

---

[53] In 2020 and 2021, firearms contributed to the deaths of more children ages 1-17 years in the U.S. than any other type of injury or illness. See McGough in References.

> *"Despite their long-standing knowledge that recycling plastic is neither technically nor economically viable, petrochemical companies. . . have engaged in fraudulent marketing and public education campaigns designed to mislead the public about the viability of plastic recycling as a solution to plastic waste."*
> "The Fraud of Plastic Recycling" report

Like cookie-thieving kids with crumbs on their mouths, plastic manufacturers flat-out misrepresented plastic recycling as a solution to the oversupply problem—that they created. It's like when I stash the pile of empty Doritos bags into the secret bin under my desk when I'm writing, as if that makes them magically disappear.

Cities and counties, in reliance on said marketing, built recycling programs and facilities, many of which are now shuttered due to *foreseeable* economic losses. Our tax dollars *not* at work, thank you so very effing much.

> *"If not for the Big Oil and the plastic industry's lies and deception, municipalities and states would not have invested in plastic recycling programs and facilities—many of which have been shut down due to foreseeable economic losses. The industry not only misled municipal and state agencies to believe that plastic recycling was a viable solution to plastic waste, but also*

discouraged them from pursuing other, more sustainable waste management strategies."
"The Fraud of Plastic Recycling" report

Not content with wasting our tax dollars, the petrochemical companies had the gall to pin the responsibility for the plastic proliferation on us, the consumers. It's our fault that our shampoo and detergent bottles aren't being recycled as promised because we don't put our discards into the right bin? Or because we don't make enough craft projects from the multitudes of yogurt cups? All I have to say about that is. . . .
XXXXXXXXXXXXXXXX.

Uneffingprintable.

Is there anything a fossil fuel company says that we can trust? Rhetorical question.

The worst of the worst news is this: the fossil fuel companies have no intention of slowing down the plastic pipeline. Their current plastic-making villainy already produces a not insubstantial 232 million tons of CO_2 gas emissions per year, equivalent to emissions from 116 coal-fired plants. Those emissions result all along the production line, from the drilling of the oil, to cracking the ethane,[54] forming the plastic, and, ultimately, from incinerating it, as some 12% of plastics are. And that's what's already happening.

The petrochemical companies are doubling down on

---

[54] Ethane is the key material for plastics. It's a compound in petroleum that's commonly isolated when refining gasoline. It's also a byproduct of fracking, so guess who wants to keep on fracking, to supply themselves with ethane to produce more plastic?

investments in plastic-making operations. They've spent more than $200 billion on more than 300 plastic and related chemical projects just in the U.S. since 2010.

"Companies like ExxonMobil, Shell, and Saudi Aramco are ramping up output of plastic — which is made from oil and gas, and their byproducts — to hedge against the possibility that a serious global response to climate change might reduce demand for their fuels."
The Plastics Pipeline report

All those new plants mean the amount of greenhouse gases produced by the U.S. plastics industry will skyrocket, exceeding the emissions from coal-fired power plants by 2030. When the plants currently being permitted or built go online, they will release as much greenhouse gases as 27 more coal plants.

The sickening irony is that this is their backup plan. As the world pushes to reduce fossil-fuel powered energy to combat climate change, petrochemical companies are ramping up their production of plastic, to make up for their losses. Yes, they're using their fossil fuel reserves, which they can't sell because of climate change concerns, as the raw material and the energy source for manufacturing more and more plastics, which in turn generates more greenhouse gases.

MAKE THAT MAKE SENSE.

Does driving an electric vehicle, taking the bus or

carpooling even matter if the companies are just inventing ways to burn their fossil fuels, anyway?

Excuse me while I shriek-cry into the brown bag I keep near my desk just for this purpose. My paltry efforts to avoid single-use plastic now seem pathetic. What good does it do to take my own bamboo fork out to dinner if the ExxonMobils of the world intend to double the current production of plastics—and associated greenhouse gas emissions—when we are already swimming in too much? Where are the consequences, the repercussions, for ruining every environment, invading the body of every living creature, and compromising every life support system on Earth?

After everything I wrote in *Love Earth Now* about making thoughtful choices, about making peace with doing the best we can, given the choices we have . . .

I am shattered.

Looks like I picked the wrong lifetime to quit smoking. Drinking. Sniffing glue.[55]

I need a minute.

.
.
.
.
.

And that's just an overview of our *plastic* "novel entities" problem—from an altitude of outer space. This

---

[55] Yes, that's a nod to the 1980 movie, *Airplane*, where the stressed-out air traffic controller makes a series of "I picked the wrong day to quit ____" and mentions a series of substances.

category in the "Earth Out of Bounds" report also includes synthetic chemicals and nuclear waste. But I'm too far gone at this point. I can't even fathom the additional devastation wreaked by the 40,000 or 350,000[56] synthetic chemicals in use today, let alone the fallout of all our nuclear waste. I'm gonna need a mind-numbing gummy just to finish writing this book.

Suffice to say that it's long past time, IMHO, for Big Oil to pay for their fraud, much as we have (eventually) demanded that tobacco and opioid companies be held accountable for pushing their own versions of dangerous products. Take some of the billions you've made on selling us a plastic bill of non-recyclable goods, petrochemical companies, and start cleaning up the mess you've made.

If toddlers can learn to clean up their spilled milk messes, so can you.

NEWS BREAK

This just in. On September 24, 2024, the State of California sued ExxonMobil, alleging that the company waged a "campaign of deception" to mislead consumers into believing that recycling was a viable solution for plastic waste. Believe me when I tell you that I am covered in goosebumps. Have my rants finally paid off?

"Exxon Mobil knew that 95% of the plastic in the blue bin was going to be incinerated, go into the environment or

---

[56] Estimates vary widely.

> *go into a landfill. They knew and they lied."*
> *California Attorney General Rob Bonta*
>
> This is going to be interesting. Watch this space.

So what song am I singing here? I got my cynical, "does this even matter?" self to drive the extra mile to shop at the "refill" grocery store, instead of loading up my shopping cart with plastic-encased products at the nearby market. I remembered to take my reusable glass jars and drawstring bags to fill with liquid dish soap, crackers, dried fruit, nuts and rice. Please, hold your applause.

I also signed up for the Ridwell service that picks up some of the plastics that the city doesn't take, like plastic film and multi-layer plastic bags. Despite my best intensions to avoid them, there's still a steady stream because my family does eat cheese, granola bars, crackers, and too many chips. Please don't tell my cardiologist.

As glad as I am for this service, it's not a perfect fix. Even though the city already sends out trucks to pick up trash and recycling, I have to pay someone else to drive another fossil-fuel burning, greenhouse-gas-emitting vehicle[57] on the traffic-clogged streets of L.A. to pick up these unwanted plastics, then transport them to a sorting facility, and ultimately, on to the remanufacturer.

That said, I'm thrilled that my family's too many

---

[57] Ridwell does partner with 3Degrees to offset the company's carbon footprint.

potato chip bags get remanufactured by Hydroblox, Inc., using no glue, binders or chemicals, into industrial drainage solutions. Thwarting the diabolic plot of the petrochemical companies to ramp up production of virgin plastic pleases me to no end. I just wish we weren't racking up so many impacts to make this good thing happen. Ridwell did advise that they partner with 3Degrees, a "pioneer in climate solutions," to offset the company's carbon footprint.

I also recognize that my plastic-turned-Hydroblox will someday become permanent waste, that even this good thing is temporary. But I dream of a day when someone figures out a more permanent solution, a day when robot WALL-E[58] comes along to transform all our noxious, non-compostable waste. It's a fantasy that I indulge, just to make it through the day.

It's the song I have to sing today.

"A bird does not sing because it has an answer. It sings because it has a song."
~Chinese proverb

---

[58] Reference to the 2008 Pixar movie, *WALL-E*, featuring a lonely robot left to clean up the garbage on Planet Earth while her residents cruise space in luxury starliners—having learned nothing about their wasteful ways.

RIDWELL

Despite my handwringing over the emissions, Ridwell provides much-needed services coupled with excellent customer service. In addition to picking up plastic film (like those flimsy grocery bags) and multi-layer plastics (my chip bags), Ridwell also collects old batteries and light bulbs, for safe disposal. Note that categories of items picked up vary by region.

Yet another benefit is the "featured category," which inspires me to clean out my drawers and closets. Each pickup, they partner with a local resource, giving members an easy way to donate reusable handbags and bras to a mobile clothing closet, art supplies for a children's creativity lab, and coats for unhoused youth.

They currently serve the Seattle, Portland, Austin, Atlanta, Denver, Los Angeles, San Francisco Bay Area, and Minneapolis/St. Paul metro areas. They are expanding quickly, though, so check http://www.ridwell.com/plans to see if they are picking up in your neighborhood yet.

http://www.ridwell.com/

BEYOND PLASTICS MOVEMENT

If you're as sickened by the despicable "recyclable plastics" deception campaign as I am, check out the

Beyond Plastics project. The mission is straightforward—and ambitious:

"Our mission is to end plastic pollution everywhere."

https://www.beyondplastics.org/

Founded in January 2019 at Bennington College in Bennington, Vermont, Beyond Plastics combines the expertise of environmental policy experts with the passion of grassroots activists to effect change. Beyond Plastics recognizes that eliminating plastic waste requires more than us consumers refilling our water bottles.

> *"It will take changes at every level of our economy and civil life to stem the tide of plastic pollution."*
> BeyondPlastics.org

Beyond Plastics seeks to educate media, policymakers and consumers on the plastics crisis, to urge businesses to eliminate single-use plastics, and also to help block new plastic manufacturing facilities.

No small goals. They need your help.

With more than 100 local groups across the U.S., there may be a local group near you. If not, start your own, using the grassroots organizing training, tools and support offered by the Beyond Plastics organization.

Oh, and if you eat plastic-wrapped cheese, check out their "Cheese Packaged in Plastic May Expose You to Harmful Chemicals" page. Yikes.

https://www.beyondplastics.org/act

HABITS OF WASTE is one of those many groups associated with the Beyond Plastics movement. Founded by Sheila Morovati, she was first inspired to act by her young daughter, a particularly persnickety eater who would eat only the same dish from the same restaurant. The chosen restaurant offered "free" crayons with every kid's meal, and the wastefulness irked Morovati. She convinced restaurants to collect their used-once crayons to donate to schools and Head Start programs, saving over 20 million from the landfill to date.

She went on to spearhead a ban on plastic straws and cutlery in Malibu, then launched the #CutOutCutlery campaign to convince Uber Eats and other food delivery companies to make plastic cutlery available only on request.

Morovati, who immigrated to the U.S. as a child, sees and calls out our wasteful habits. "It's so embedded in our daily routine, it's so normal that we don't even see it," she says. She brings our attention to the insanity of our throwaway society, then devises campaigns to direct action. "Focus on solutions, not complaints," she advises. Habits of Waste creates effective grassroots campaigns that are simple and convenient enough for even the busiest of us to do. Current campaigns include:

- **Bars over Bottles**: call for manufacturers to create bar forms of hygiene products like shampoo and conditioner.

- **rethink Tap**: educate people about the good quality of tap water supplies (for most of us) and discourage the use of water bottled in plastic (which likely contains nanoplastics).

- **Ship Naked / Ship Greener**: encourage Amazon, Walmart, Target and other big etailers to ship boxed products without an additional mailing box ("ship naked") and to use the correct size box to eliminate unnecessary packing materials.

- **Lights, Camera, Plastic?** urge the motion picture industry to reduce the depiction use of single-use plastic in their productions and replace them with reusable utensils and food containers. Because monkey see...remember when people used to smoke cigarettes in every movie? Okay, maybe you're younger, but that used to be a thing.

Each campaign includes clear and easy steps for individuals willing to participate. Check out the Habits of Waste website for details. https://habitsofwaste.org/

SEVEN
My Heart Is This Bamboo Fork

"This bamboo fork represents my heart."

That's what I wish I'd said instead of blathering on like the raving lunatic that this new acquaintance surely believes I am. Why, oh, why, do the perfect words always come well after I've mortified myself?

Husband and I had gathered with a few friends and a new acquaintance at a local restaurant, and I was giddy, going OUT to eat still an exciting adventure in those early post-pandemic times. Finally, I could eat something I did not cook, chop, slice, dice, or nuke!

As we exchanged pleasantries, I pulled out my bamboo utensils, in eager anticipation of the meal ahead. Though restaurants were open again, many, like this one, still provided only plastic utensils. Don't get me started.

Determined not to let the plastic debacle ruin my Night Out, I kept my focus on my own utensils when I heard an acquaintance at the table ask:

"You brought your own.....fork?"

My moment of Zen faded as the internal debate

regarding the appropriate response ensued among Me, Myself and I.

"Was that a tone of snark I detected?" I asked Myself.

"Surely not, he's just curious!" Myself replied.

I remain unconvinced, snarling, "That look on his face is definitely a smirk!"

My face now beet red, my introvert self summoned the will to defend my choice of cutlery. Eyes narrowed, I turned to face the new acquaintance, someone not yet apprised of my eco-neuroses.

"Um, yes..." I said aloud to my companion. "This reusable fork may look like a stupid thing to you. And...it's just a fork, as you say. Before we go further in this conversation, however, there's something I need you to know about me, seeing as how we recently met.

"This fork is what stands between me and utter insanity. You see, my life's work, as it turns out, is honoring and thanking Earth for giving birth to, nurturing and loving me and my beloveds. That's one of my mantras I repeat all the time.

"Facing the Bad and Worse news for Life on Earth is the most gutting challenge I face each day. And my family counts on me to make it through. For my own sanity, I must believe that the Small Things I Do Matter. That bringing a fork or a straw, one I can take home and wash for reuse, instead of mindlessly tossing a plastic one, means I'm doing my work, what I came here to do. Because if I don't have that....I am sinking so deep into the mind-numbing gummies that I'm of no use to anyone." I blinked furiously, forcing back hot tears.

I need to work on the ending. Obviously. Because

MY HEART IS THIS BAMBOO FORK

nothing screams "snowflake liberal" like public sobbing over a simple fork. But dammit, that's where I am right now.

Not that I think of myself as a snowflake at all. It takes grit just to be conscious in this world, which is seemingly hellbent on numbing us to the ongoing destruction of our essential life support systems just to make someone else stupid-rich. Pay no attention to the billionaire laughing behind the wall of smoke streaming from a fossil-fuel-powered monstrosity.

I recently watched the IMAX film *Hubble 3D*, a documentary full of rich images of our Universe, where the madness of our Nature-destroying ways was writ large. Billions upon billions of stars[59] twinkle across the vast Universe, many orbited by planets of all sorts and sizes....and yet we've only discovered one that supports Life as We Know It.

We've been given a priceless gem. Are we really going to toss it into the garbage can because we love money more? Because caring for the planet is "woke"? Because we're too busy? Because "God will provide"?[60]

As all this churns in my brain, I hear my dinner companion reply.

"That's cool. I've just never seen a bamboo fork before."

---

[59] Between 100 and 400 billion stars and 100 billion planets in the Milky Way galaxy alone.
[60] In my opinion, God has already provided. We're the snotty toddlers smashing the gift and demanding a new one.

"I long to accomplish a great and noble task, but it is my chief duty to accomplish small tasks as if they were great and noble."

Helen Keller, activist for disability rights, women's suffrage, labor and pacifism

SOAPBOX
THE BIOPLASTIC CONUNDRUM

Have you ever stood in the grocery aisle, on the eve of a party, trying to figure out if there is ANY sustainable way to avoid washing dishes? Me, too, my friend, me, too. Those beguiling bioplastic wares in vibrant colors tempt me, but when I put them into my compost bin, they came out looking the same, as neon green as when they went in. What's the deal?

If you read the fine fine print on the boxes (you may even have to check the website), many say they are compostable only in a commercial facility. Backyard bins don't usually generate sufficient heat.

And if you're fortunate enough to have a commercial composter near you—or your municipality delivers your compostable waste to one—many of them don't want the stuff like bioplastic utensils or biobags, for both logistical and legislative reasons.

The similarity in appearance to plastics, a great selling feature, perhaps, is another part of the problem. It's impossible for the compost facility to know whether those plastic-looking utensils are made from petroleum or plants. Check with your commercial compost facility before sending them in.

Cities in California are under a mandate to reduce compostable wastes sent to landfills by 75% by 2025. Los Angeles now collects food scraps, pizza boxes, coffee filters, bamboo, untreated wood and corks in the same bin as green yard waste. They don't accept, however, manufactured products like "certified compostable" tableware and biobags. Nor does the area's largest private hauler. The key issue seems to be they take too long to break down to be compatible with commercial composting practices.

As for the corn-based cutlery and cups, do we really need to be devoting precious acreage to growing crops for an item to be used once and tossed? Is that the best use of precious farmlands, given the time, water, acreage, and increasing challenges of growing food in a warming world with more unpredictable rainfall? I know my answer.

Compare to tableware that's made from bamboo and sugar cane, which are made from shucks and other "waste products" of plants already being grown for other purposes.

So, what is an Earth-loving party host to do?

Yes, reusable is always best—even Martha Stewart says it's fashionable to mix and match thrifted tableware of different designs. But what if there just aren't enough hands for all that dishwashing?

Use tableware from fibrous or edible sources whenever you can, like these:

1. Fiber-based plates: paper, bamboo, bagasse (sugar cane stalks), palm leaves, etc. These must be the plain Jane variety, no brilliant-color designs which are likely printed on a plastic coating.
2. Utensils that are uncoated bamboo, wood or other 100% plant-fiber based.
3. Stir sticks made of wood or an edible option like pasta.
4. Straws made of 100% uncoated paper, plant fiber or edible sources.

The bottom line for me is this: if it looks like it was once part of a plant or tree, it's probably compostable. If it resembles plastic, run as fast as you can. And if it's truly compostable but wrapped in plastic? Drop to the floor of the grocery store aisle and cry.

But that's just me.

"It isn't easy being green."

Kermit the Frog

EIGHT
Playlist Track 3: Dead Zones

♩ **Recommended Musical Pairing** ♩
"Whose Garden Was This?"
~Tom Paxton

Another already-exceeded measure in the "Earth Out of Bounds" study is "biogeochemical flow." What's that, you ask? I didn't know either. Turns out, it basically means we're messing with Earth's nitrogen and phosphorous cycles, mainly because of our use of fertilizers and other chemicals in agricultural operations.[61]

The fertilizers that farmers and lawn enthusiasts most commonly apply contain a lot of nitrogen and phosphorous. When rains fall or the irrigation kicks on, much of that fertilizer gets washed away, draining into lakes, estuaries, aquaculture ponds and coastal waters. This is good news for the algae, phytoplankton, and

---

[61] City runoff waters are also a significant factor.

seaweeds that gobble them up much like a ravenous glutton at an all-you-can-eat buffet. When these organisms stuff their faces[62] at the fertilizer buffet, they grow out of control on the water's surface, creating an "algal bloom." A "bloom" may sound nice, like when the nasturtiums take over my entire yard. Unfortunately, the algal blooms are more like dreaded dinner guests who not only eat all the food in your kitchen, but then proceed to starve you out.

These massive algal blooms block the sunlight and oxygen that life beneath the water surface requires for photosynthesis and respiration. Fish, shellfish, sea turtles and birds suffocate and die in masses as a result, causing a "dead zone."[63]

The biggest dead zone in the United States pops up in the Gulf of Mexico every summer, after farmers upstream treat their fields for the growing season. The Mississippi River acts like a funnel for waters draining from 31 upstream states down to the Delta, one of the most productive fishing grounds in the U.S.

How big is this dead zone? It varies by year, but the National Oceanic and Atmospheric Administration ("NOAA") says, over the past 37 years, it has averaged 5,000 square miles—the size of the state of Connecticut. It was 8,776 square miles in 2017—the size of New Jersey.

Okay, but what do those numbers mean? The Gulf of Mexico is enormous, so how do we know if that's really a

---

[62] Okay, I know they don't have faces, but I have no idea how they actually eat, so work with me here.
[63] Not to be confused with a gathering of Grateful Dead fans.

PLAYLIST TRACK 3: DEAD ZONES

problem? Meet Dean Blanchard, a shrimp distributor for more than 40 years in Grand Isle, Louisiana. Blanchard has said:

> *"Sometimes we'll get hundreds of dollars of shrimp a day, then the next day everything's gone," says Dean, "when the dead zone comes it just kills everything."*
>
> Dean Blanchard, owner of Dean Blanchard Seafood

As if dealing with oil spills[64] and super-sized hurricanes[65] wasn't enough, along come the downstream pollutants that kill your catch. Blanchard hopes to pass on the business to his nephew. I pray there is a business to pass on.

These not-so-pretty blooms impact not only fisheries but ruin recreation areas and make humans sick from respiratory irritations and bacterial infections. Who wants to come home from a relaxing day at the beach and throw up?[66]

And they're costly. The Environmental Working Group estimated in 2020 that communities across the U.S. have spent more than a billion dollars to deal with toxic algae outbreaks over the prior ten years. That

---

[64] In 2020, Blanchard said his business is 15% of what it was before the British Petroleum 4.9-million-gallon oil spill in 2010.
[65] Hurricane Ida did $1,000,000 in damage to Blanchard's processing facility in 2021.
[66] And not because the potato salad went bad in your picnic basket.

amount doesn't include the downstream costs of losses in tourism, commercial fishing, or property values.

And like everything else that's already awful, climate change will make these dead zone parties more frequent. Blue-green algae only blooms in water warmer than 60 degrees Fahrenheit. Guess what will make those conditions more frequent and persist longer?[67]

So what song am I singing here? There's not a lot I can do about agriculture; I had to give up my small vegetable garden after it became a party palace for pests. I still shudder to think of the worm-infused cauliflower I once grew. Much as I respect and honor all Life, I don't want it all in my food.

My failures make me appreciate the hard work of brave souls who produce our food even more. I wrote at length about how much I honor the courage and tenacity of farmers in *Love Earth Now*.

"Farmers stare down the vagaries of nature, pests, disease, commerce, demand, price controls, regulations, and—let us not forget— the latest 'let's-all-eat-kale!' fads to grow our food, all the while knowing that months of labor and all their capital, hopes, and dreams can be wiped out by a single pest."

Love Earth Now

---

[67] I know you've figured this out, but for the fact checkers, the answer indeed is a warming planet.

Our family gets our fruits and veggies through a community-supported agriculture (CSA) service, so we can buy from farmers who don't apply any synthetic fertilizers or bee-killing pesticides. I'm grateful to be able to support them directly, especially those doing their Hoover dam-dest to feed us while also protecting their workers and the ecosystems where they farm.

FIND YOUR CSA

Do you have a CSA near you? LocalHarvest maintains a comprehensive list of them, as well as tips on choosing a CSA, so you can search out one that fits your needs. It's a great way to get farm-fresh food for your family, and to support hardworking farmers by giving them a steady roster of customers.

https://www.localharvest.org/csa/

I'm heartened to read about caring Midwestern farmers like Tim Little of Minnesota. Tim and his farming friends formed a collective to find ways to solve the downstream pollution problem. Tim started planting "cover crops" across his 2,000-acre farm to reduce fertilizer use and soil erosion and improve soil health. Cover crops are usually grown after the main crop is harvested or in fallow periods to enrich the soil with nitrogen and essential nutrients and reduce the amount of fertilizer needed. They also reduce soil and moisture losses, suppress weeds and pests, and provide habitat for beneficial insects. So, raise your glass, please. Three cheers for cover crops!

It's still a delicate balance, determining the amount of fertilizer to apply, since heavy rains wash so much of it away. Guess what Midwestern farmers are seeing a lot more of, thanks to climate change? If you said "heavy rains," I know you're paying attention.[68]

I applaud upstream farmers caring about what happens in the downstream realm, farmers who take responsibility for what they put into the environment—unlike some truly awful plastic manufacturers I could name. May even more farmers adopt less-polluting ways of feeding us all—and may we thank them every chance we get.

FUTURE FARMERS OF AMERICA

The Future Farmers of America (FFA) organization educates some 950,000 students about all aspects of farming, including "climate smart agriculture." The changing climate impacts everything from the length of the growing season to increased variability in rainfall. I'm glad this organization of some 9,000 chapters across the United States, Puerto Rico and the U.S. Virgin Islands, provides up-to-date intel. These young farmers of tomorrow will face challenges their predecessors never imagined.

FFA trains young people for careers in, not just farming, but also chemists, veterinarians, government officials,

---

[68] And if you didn't, I'm guessing you won't even bother to read this footnote.

entrepreneurs, bankers, international business leaders, and teachers. FFA also teaches young people the personal development and leadership skills to succeed in whatever their chosen field.

Is there a chapter near you? I did a search and found one at the local high school, here in urban Los Angeles.

https://www.ffa.org/about/

4-H

4-H is the largest youth development organization in the United States, educating some six million young people across the country. The "4-H" name represents the key values of Head, Heart, Hands, and Health. 4-H offers programs to develop skills in leadership, citizenship, and life skills, for application in the fields of health, science, agriculture and civic engagement.

4-H also provides essential climate science education. The Climate Science curriculum includes lessons about the causes of, evidence for and impacts of climate change. Members are encouraged to engage in environmental projects, like improving natural habitats. The more time young people spend in nature, the more they care about it. I'm glad for the 4-H organization getting them out there and learning.

The programs are available in every county and parish in the U.S. https://4-h.org/

NINE
Cooking Up Therapy: Recipe For Resilience

Living in our ever-changing world drives many of us to search for new therapies for surviving the chaos—therapies that are kind to the liver. So many of us, in fact, that I'm surprised we haven't broken the Internet. Yet. If only there was a recipe for navigating a world that seems hellbent on making us feel like drunk landlubbers on a galleon in a tempest. How many times have I wished upon that star!

As I witness the norms I once believed inviolable fall away like weight-loss resolutions on Valentine's Day, I've found myself longing for the warmth and comfort of my grandmother's kitchen. That's where my sister and I could count on someone cooking up something scrumptious because that sure wasn't happening at home. Our mom hated to cook, and she did her best to dissuade us from expecting her to produce anything edible.

But in Grandma's kitchen, only Good Things happened. She seemed to have a recipe to cure any ailment. Fell down and skinned your knee? There's a slice of lemon meringue pie for that. Lost your favorite toy? Let's bake an angel food cake together. Boy broke your heart? Let's eat the leftover homemade peach ice cream straight from the bucket.

Grandmother isn't around to cook my broken heart well anymore, I'm sorry to say. But I still have her recipes. And I got to thinking... maybe there is an adaptation I can make, one that preserves the goodness while soothing the hurts of a new and terrible time.

I dusted off Grandmother's cookbook stuffed full of yellowed index cards with family-tested recipes written in her elegant handwriting—the good old ones that didn't suck out the enjoyment by eliminating carbs, salt, fat, and everything delicious. Why start from scratch when I can rely on a surefire satisfier? Lucky you, I've decided to share this sweet therapy.

GRANDMA'S BROWNIES

Makes 1 or 8 servings, depending on what you consider a "serving."

INGREDIENTS

- **2 squares unsweetened chocolate.** Make it Fair Trade if you can swing it. Chocolate produced from cacao beans farmed by enslaved children[69] is not so sweet.

---

[69] Some 1.5 million children perform hazardous tasks, from handing chemicals to wielding sharp tools, on the cacao farms in Cote d'Ivoire and Ghana which produce 60% of the world's cacao beans.

- **¼ pound butter** or, if you're vegan (or your cardiologist insists), 1/3 cup vegetable oil. If you're using cannabutter, because you need a little *extra* comfort, you'll have to Google that recipe; it's not in Grandma's cookbook.
- **1 cup granulated sugar**, or, to reduce your food-mile footprint,[70] ¾ cup local honey.
- **2 farm-fresh eggs,** from your backyard, if you're a hen host, for the ultimate low food-mile score.
- **½ tsp vanilla extract** if your budget allows[71]
- **1 cup chopped walnuts,** if you're not allergic
- **¼ cup all-purpose flour**
- **¼ tsp salt**

STEPS:

Preheat oven to 325°F. Please do not heat up the oven and forget what you are doing, burning up gas or electricity and fomenting climate change while you try to remember why you're in the kitchen. Been there too

---

[70] "Food miles" represent the distance your food travels before it gets to you, burning fossil fuels and racking up emissions all along the way. Sugarcane grown in Hawaii gets shipped to a sugar refinery in San Francisco, then to New York to be packaged, and then to grocery stores across the country . . . even those in Hawaii. A packet of sugar in Honolulu may have traveled 10,000 miles. Make that make sense.

[71] If your pantry bottle lists "artificial vanilla flavoring," it's likely made of guaiacol which comes from petrochemicals. If it says it's "vanilla extract," use it sparingly because it's pricey these days, and for good reason. Vanilla extract is made from the vanilla bean orchid which only grows in a few places in the world, and it must be hand pollinated. Some 80% of the world's vanilla extract supply comes from Madagascar, a country hit hard by drought and multiple hurricanes, thanks to climate change (four cyclones hit Madagascar in one month in early 2022). But good news! You can grow your own vanilla orchid if you have a good 8 to 10 feet available in your greenhouse. . . and a lot of patience.

many times, walking into the kitchen wondering why in blazes it's so hot in there and there's nothing to eat.

Melt chocolate and butter (or your chosen substitute) in heavy skillet. Remove from heat and stir in sugar or honey.

Beat in eggs with all the fervor of someone (possibly you) who's furious at the awful choices we've been given by fossil fuel companies who lay responsibility for cleaning up the climate and plastics messes on us bewildered consumers.

Quickly stir in vanilla, walnuts, flour and salt before you can get too worked up about the plastic disaster to finish the brownies.

Spread batter in a well-greased 8x8 pan—or whatever you find that's relatively clean. Eco-anxiety leaves little patience for washing dishes. Or maybe that's just an excuse I made up.

Bake for 40 to 95 minutes, depending on whether you lose track of time. If you forgot to turn on the timer, remove pan from oven when the fire alarm goes off. Little tip I learned from my kitchen-hating mother.

Please do remember to turn off the effing oven when you remove the pan. Our climate rests in ~~your~~ all our hands.

Allow to stand for ten full minutes before shouting at whoever's handy to come cut this masterpiece while you go out for a smoke. Or to smack a box on the recycle bin. Too much time in the kitchen kills brain cells. Another helpful tip learned from Mom.

Or, if your anxiety refuses to share, skip the cooling and cutting steps entirely. Grab a large spoon. Dig in.

Cover any remaining brownies, if you possess the willpower (and goodwill) to leave them for others. Please do consider the awful environmental impacts of plastic wrap when you choose your covering, unless that's all you have, then go for it because no one needs another guilt trip right now.

Above all, remember that cooking is meant to be a joy, so if this recipe doesn't spark any joy, then decluttering expert Marie Kondo says you should toss it.

SOCIETY OF FEARLESS GRANDMOTHERS OF SANTA BARBARA

These grannies aren't just cooking up brownies to console themselves. They take action, from blocking off streets to protest to displaying "No New Fossil Fuels" banners in front of offshore oil wells. Ranging in age from 60 to 90, these climate-caring women keep up to date on legislation, attend local government meetings, knock on doors, and write a lot of postcards. They've had knee replacements, hip replacements, and broken wrists, and still, they keep going.

They also employ craftivism, a form of activism utilizing the domestic arts, such as stitching and knitting, to express messages of protest which dates to the days of the suffragettes (and beyond). For Earth Day 2020, they created a collage of images of their grandchildren holding letters that spelled out messages to the governor and other elected officials. In 2022, they made a quilt to illustrate solutions to the climate crisis which they display at various venues to spark conversations and encourage activism.

Founder Irene Cooke learned about the perils of global warming from her niece, a climate scientist. Being a mother and grandmother, she knew she had to do something to protect the next generations. She completed Al Gore's Climate Reality Project training and began speaking to groups. When she relocated to Santa Barbara, California, Irene didn't allow her lack of location connections to stop her.

Inspired by indigenous women in the Bay Area, including Alison Ehara Brown and Pennie Opal Plant of the original Society of Fearless Grandmothers, she was determined to gather a group of kindred spirits in Santa Barbara. She knocked on the doors of the local Unitarian Universalist churches, the Santa Barbara chapter of 350.org, Women's March, and others. The Santa Barbara branch of the Society of Fearless Grandmothers was born thanks to her determination.

Irene notes that the Society works closely with the California offices of the Center for Biological Diversity, a national organization that organizes advocacy and awareness events, provides resources and coordinates efforts of participants. The Society has also collaborated with international groups such as 350.org, Third Act, Fridays for Future and Stop the Money Pipeline. Coordinating their actions with larger organizations connects the Society members with the greater activist community, while still allowing the flexibility to tailor their efforts to local issues.

https://www.fearlessgrandmotherssb.org/

"We have the solutions to the climate crisis—NO miracles are needed! But we are running out of time: We need the political will to implement existing solutions in the next five years before critical climate tipping points are reached."

Irene Cooke, founder of the Santa Barbara chapter of the Society of Fearless Grandmothers

---

TEN
Playlist Track 4: Deforestation

♫ **Recommended Musical Pairing** ♫
"Heart of Oak"
Richard Hawley

This already-exceeded measure in the "Earth Out of Bounds" report concerns changes in land use. No surprise to anyone, I'm guessing, to say that we humans have altered the natural landscape a lot. Pretty sure there were no mini-malls or gigafactories when humans first arrived on this planet. No dams, no skyscrapers, no oil refineries, no Wall Drug billboards advertising free ice water for 650 miles.[72]

The "Earth Out of Bounds" report used the conversion of forests to range, farmland, and other uses as the gauge for

---

[72] If you've traveled any part of I-90 between Minnesota and Billings, Montana, you know. Wall Drug Store of Wall, South Dakota, established in 1931, features cowboy-themed stores, restaurants, arcade, and an 80-foot brontosaurus sculpture. But it's their multitudinous, hand-painted billboards and mile markers that made them legendary.

this "land system change" boundary, given the undeniable link between forest loss and climate change.[73] It's hardly surprising considering trees absorb carbon, produce oxygen, provide cooling via evapotranspiration, and some even generate their own rainfall and weather systems. Eliminating destructive deforestation practices is essential to ward off the worst impacts of climate change, to preserve biodiversity, and to protect water and soil resources.

If you doubt the importance of trees, I suggest you have your lungs removed and see if you can still breathe. So why are we still taking out forests faster than a scoop of ice cream melts on a sidewalk in Death Valley in July? Something like a billion acres of forests were converted to other uses between 1990 and 2020. The "Earth Out of Bounds" report says that the loss of Amazon rainforest lands has been so extensive that the safe boundary for its forest health has already been surpassed. A moment of silence for what some call the lungs of the planet.

Causes of deforestation in the Amazon include agricultural expansion, mining, logging, fires, new roads exposing the deep forest and climate change. But our cheeseburger habit in the U.S. may be the biggest culprit of all.

"If the Amazon dies, it will be beef that kills it. And America will be an accomplice."
"Devouring the Rainforest," *Washington Post*

---

[73] Forests are also critical for the biodiversity metric, providing habitat for 80% of amphibians, 75% of birds, and 68% of mammals.

The United States bought more than 320 million pounds of Brazilian beef in 2022, and that number is expected to increase. The world's biggest beef supplier, JBS, has been repeatedly accused of buying cattle raised on illegally deforested land. Despite being fined $7.5 million by the Brazilian environmental in 2017 for such practices—and claims by JBS that they are doing better— the allegations continue.

And none of that accounts for the ranchers who fly under the radar. The cattle supply chain in Brazil is so complex, involving thousands of ranches across a vast and rugged territory, that monitoring it all is near impossible. Schemes like shuffling cattle from an illegal ranch to a legal one before sale, the worst kind of switcheroo,[74] make it the ultimate shell game. We can't see the shyster or the shells when standing in line at Burger King.

Scientists keep urgent watch to figure out when all the stresses on the Amazon ultimately will cause the entirety of the Amazon rainforest to collapse. If that dark day ever arrives, the resulting carbon emissions would equal an entire years' worth of global emissions many times over (perhaps twenty times). Habitat for more than a million different species of animals, insects, plants and other forms of life—many of which are current and yet-to-be-discovered sources of our medicines—would vanish. Disastrous soil erosion, disrupted rainfall patterns

---

[74] Unlike discovering a dollar under your pillow instead of the dead tooth you placed there the night before. That's a very nice switcheroo, Tooth Fairy.

around the world—and a jillion[75] other bad things would result, especially for the couple million indigenous people who live there. How's that cheeseburger sounding now?[76]

Beef isn't the only culprit that's causing a run on forest land like shoppers lined up outside the big box store on Black Friday[77] morning. Large-scale agriculture accounts for some 40% of forest losses around the world, not just in the Amazon. Agriculture-caused forest losses are largely driven by, yes, cattle ranching, but also for raising soya beans and oil palm.[78] And a chunk of those soybeans goes to feed our food animals like beef cattle, so that's a double whammy for the cheeseburgers.

How can we eco-anxious consumers ensure that the food we buy isn't from illegally logged lands? If you're in Europe, you can rest a lot easier. As of the end of 2024, the European Union Deforestation Regulations requires suppliers of commodities to prove their products, like beef, wood, cocoa, soy, palm oil, coffee, leather, chocolate, tires and furniture, are deforestation free. May this type of law find its way to our American law books, to give us all some tree-hugging peace of mind. If we can ever stop worshipping our "economy today over environment tomorrow" sacred cow, that is.

---

[75] "Jillion" being roughly equivalent to SCADS.
[76] No judgment intended. I'm the first to acknowledge that knowing something is "bad," makes it even more appealing. Margaritas, cheese, and garlic fries being my personal cases in point.
[77] A nefarious-sounding term that nowadays means to the day when retailers go from operating at a loss ("in the red") to making a profit ("in the black"). But it started in Philadelphia in the early 1960s to describe the resulting chaos from hordes of suburbanites coming to the city to shop and attend the annual Army-Navy game. You're welcome.
[78] Other causes include small subsistence farms (33%), urban expansions (10%), infrastructure (10%) and mining (7%).

What if we just stopped buying food from companies who refuse to guarantee that no forests were clearcut to produce it? Probably easier said than done, given all the greenwashing[79] and empty promises out there.

McDonald's, Brands International (Burger King) and Yum! Brands (KFC, Pizza Hut and Taco Bell) have made commitments to eliminate purchasing beef and soy from deforested lands by 2030. Sooner would be better, of course—every year of deforestation means another loss of a treasure trove of plants, animals, medicines, carbon sinks,[80] and more. But at least there's some acknowledgement of responsibility.

On the Deforestation Wall of Shame,[81] however, you'll find these brands that have made ZERO commitments to stop sourcing their soybeans, soy milk, or beef from deforested lands.

- Subway
- Domino's
- Inspire Brands, parent of Dunkin' Donuts
- Starbuck's[82]

Of course, the fast-food companies, peddling ever more sugar and saturated fat, aren't the only ones putting pressure on precious forest resources. Many companies, governments and other institutions face calls to not only

---

[79] Looking at you, Amazon, including beef jerky on your "Climate Pledge Friendly" list. Was that a joke? Beef jerky has been named the most carbon-intensive food product, but Amazon said it was "climate-friendly" because some air had been removed from its packaging.
[80] A carbon sink absorbs and stores more carbon that it releases.
[81] I totally just made that up.
[82] Only soy is applicable here

stop supporting deforested products, but to restore deforested lands. The Bonn Challenge, launched in Germany in 2011, invited the world to restore 350 million hectares (3.5 million square miles) of degraded and deforested land by 2030. The challenge means more than just planting trees, but restoring the forest ecosystem, protecting wildlife reserves, establishing ecological corridors, riverside plantings to protect waterways, and more.

Did the United States make a pledge? Yes, the U.S. pledged to restore some 15 million hectares (about 58,000 square miles), which sounds magnanimous until you consider that's just about 1.5% of U.S. land.

But giving credit where it is due, the U.S. Forest Service has exceeded that goal, reporting 17 million hectares of forest land restored as of June 2019. Let's celebrate! We need all the dopamine hits we can get (legally). After all, 58,000 square miles—about the size of Illinois—is no small feat. Break out the champagne, gummies or whatever your personal Forest Restoration Celebration regimen allows.

Now, may we have some more, please? Because the current rate of forest devastation caused by climate change fueled wildfires ups the ante, meaning we are going to need to do SCADS more restoration.

So what song am I singing here?

Our family doesn't eat much beef due to heart disease issues, along with environmental concerns. When we do, however, we support sustainable suppliers of beef raised in the U.S. by ranchers who care about the land they manage. It may cost more, but that's another benefit of eating beef less often: we can afford to pay a little more on the few occasions we indulge.

Of course, the cheapest hack is not to eat beef at all. Not only is it better for your heart health, but it frees up acres of land required—not only for cattle grazing, but also to grow the food for the cattle. That's land we can use to grow food for humans, for solar panel fields, and for constructing cornfield baseball stadiums where baseball legend ghosts can play.[83] It also eliminates the methane[84] emissions of the gassy bovines—most of which come from their burps, not their back ends, by the way. Seeing as how each cow burps 30-50 gallons a day, and there are more than a billion cows in the world today. . . that's a lotta gas.

> *"We know meat is a leading cause of climate change. Meat alone can account for as much as 20% of global emissions. . . The average American right now eats six times the recommended amount of red meat. Six times. If we were to just halve that, it would have massive benefits for the climate, for food security and for health."*
> Sarah Lake, CEO of the Tilt Collective

---

[83] See the movie *Field of Dreams*.
[84] Not an insignificant consideration, since methane is a far more potent greenhouse gas than carbon dioxide.

> **TRIGGER WARNING FOR VEGANS**: yes, I realize there are a host of health, environmental and compassionate reasons not to eat cows, pigs, or any other animal. But I'm still gonna go there because we just can't afford to close off any essential conversations at this critical juncture. You may wish to skip over the next two pages.

If meat is a must-have in your household but you want to reduce your eco-impact, consider adopting a Meatless Monday or even substituting chicken for beef. A 2023 study by the Stanford School of Medicine found that making simple substitutions reduces greenhouse gas emissions significantly. A ground beef patty in your burger has a carbon footprint that is 8-10 times higher than a ground chicken patty and 20 times more than a vegetarian. Replacing cow milk with non-dairy milk and replacing juice with whole fruit also result in lower impacts.

#8MEALS
There's an app for that. If you want support in exploring more plant-based meals, check out the #8meals app by Habits of Waste for suggestions. The goal is to make 8 meals out of the week's 21[85] plant-based, which means you're reducing your animal product intake by about

---

[85] Assuming you eat 3 meals a day, instead of grazing all day like I do.

> 40%. Breakfast, lunch or dinner, mix them up as you like, the app offers recipes for all the meals that you and any picky eaters in your household wish to sample. Find #8meals in the App Store or "Habits of Waste" in Google Play.

When you do eat beef, consider supporting a sustainable supplier if your budget allows it. Not only will you invest in a meat that's free of deforestation woes, it's also likely to be lower in fat and higher in heart-healthy omega-3 fatty acids. Added bonus: you'll be supporting a small farm or ranch, instead of a factory farm. With all the pressures small farms face, especially in these days of a changing climate, they need our support more than ever.

You can find a list of small producers of beef[86] and lamb that the US. Department of Agriculture has certified as being grass-fed.[87]

https://www.ams.usda.gov/services/auditing/grass-fed-SVS

Here are a few other producers I discovered that are larger, but also care for their animals and soils with care:[88]

---

[86] Producers on this list market 49 cattle or 99 ewes or fewer each year.
[87] USDA also certifies farms for "responsible antibiotic use," animal welfare, hormone-free cattle and more.
[88] Provided for illustration purposes only. I don't mean to endorse any of them, nor do I have any affiliation. My meager social media presence doesn't earn me a dime in sponsorships, I assure you.

- TETON WATERS, Denver, Colorado offers 100% grassfed, grass-finished beef from ranches that practice regenerative agriculture.
- CREAM CO., Oakland, California sells beef from sustainable ranches focused on building healthy soils, conserving water and minimizing polluting runoff.
- CREEKSTONE FARMS, Arkansas City, Kansas. Their ranchers focus on humane care of the animals, from how they're raised to how they are processed.
- KELLER MEATS, Litchfield, Ohio. Keller offers humanely raised, grassfed meats, raised by farmers dedicated to regenerative, sustainable farming.

If you want to avoid burgers from rainforest-deforested land, commit to eating meat from animals raised in-country (not Brazil, for example)—if you can figure out the source of your food. The U.S. Country of Origin Labeling Act doesn't require the source country of beef to be noted on the label—even though it's required for lamb, goat and chicken. Illegally raised cows must have a better-financed lobby, or maybe I'm overly cynical.[89]

End of meat-related comments. **Vegans, it's safe to join back in here.**

---

[89] She said sarcastically.

I'm also singing for local trees. Trite as it might be for a tree hugger to say . . . I'm a big fan of trees—a true tree zealot, as anyone else living in smoggy Los Angeles ought to be. I've served on the board of **North East Trees of Los Angeles**, a local organization that plants and maintains urban trees in the nature-poorest parts of Los Angeles County—just where they're most needed to help cool the heat of the city, process some of the air pollutants from too-many cars and factories, and produce the oxygen we all need to breathe.

All good stuff, but it's the founding story that sold me. Scott Wilson, a local high school horticulture teacher and certified arborist, vowed to plant five trees a day upon his retirement. He enticed former students, family members and neighbors to help. The vow of one person is now a thriving nonprofit, one that not only plants and maintains scads of trees. That nonprofit is also developing the environmental leaders of tomorrow by employing young people from the communities served, instilling a greater sense of community pride, as well as training in urban forestry, park design-build, and environmental justice issues.

"I have always been fascinated with nature and animals. When I found out about the work that North East Trees does and the neighborhoods that they serve, I fell in love. One of those communities is Ramona Gardens, which is where my family's roots originated after immigrating from Mexico. NET recently re-landscaped a park where my

grandparents first met. To be able to make a positive impact on the community that grew my family is the best reward.
Since joining the team, I've learned so much about our Southern California ecosystem, and I'm glad for the opportunity to educate others. To me, education is the most important part of our work."
Will Guerrero, North East Trees Urban Forestry Manager and Certified Arborist

---

Let's refocus our telescope to a wider scope and give some well-deserved attention to the **Green Belt Movement** in Africa. Fueled by concerns of rural Kenyan women about freshwater streams drying up, lack of firewood for fuel, and insecure food supplies, Professor Wangari Maathai founded the grassroots movement, the Green Belt, in 1977. She established a program for individual communities to collect seeds, grow and maintain trees, both for the environmental benefits but also for empowerment of women and the eradication of poverty. Now a network of 4,000 communities, members have planted over 51 million trees in Kenya, while the movement has spread across Africa.

People like Scott Wilson and Wangari Maathai who see a need and respond, without fanfare or glorifying themselves, inspire me. How about you?

The work of the Green Belt Movement stands as a testament to the power of grassroots organizing, proof that one person's simple idea—that a community should come together to plant trees— can make a difference, first in one village, then in one nation, and now across Africa. . . . As [founder Professor Maathai] told the world, "We must not tire, we must not give up, we must persist."

U.S. President Barack Obama

FOREST BATHING

Trees aren't just good for carbon sequestration and cooling hot cities. They provide a myriad of therapeutic benefits, even more than an hour of funny cat videos. "Forest bathing," a leisurely, mindful stroll in nature, reduces blood pressure and stress, increases energy levels, accelerates recovery from illness, improves mood, and results in better sleep.

Sign me up for all of those.

Certified Nature and Forest Therapy Guide and Trainer Jackie Kuang explains:

"Forest therapy walks are not hikes or a simple wander in the woods. These walks follow a standard sequence. Each walk begins with establishing embodied contact with the present moment and place.

Next come a series of connective invitations, often improvised in the moment and adapted to the needs of participants. These may be followed by wander time and/or sit spot. The walks end with a ceremony of sharing tea made from foraged local plants and snacks. An entire walk is no more than a 1/2-mile distance. In that short distance, most people experience contact with nature in a much deeper way than they ever have prior to the walk.

On Forest Therapy walks, people have a wide range of experiences, some of which they feel are significant, even profound. Guides are trained in the skills and perspectives needed to be supportive witnesses of these experiences."

Take a mindful walk in nature on your own or find a forest bathing guide here. https://anft.earth/listings/ We need all the healthy therapies we can get if we are to navigate the many crises we face with some sort of composure. Not that ranting and raving doesn't have its place—as I'm sure you've gathered if you've stuck with me this far—but my sanity requires a number of ports in the shitstorms. I'm guessing yours might, too.

PLAYLIST TRACK 4: DEFORESTATION

"God has cared for these trees, saved them from drought, disease, avalanches, and a thousand tempests and floods. Even so, God cannot save them from fools."
— **John Muir**

ELEVEN
The Hate List

Hate. It's a loaded word, so full of karmic blowback that I never, ever use it. Except when I do.

I extracted it from my vocabulary like a Lyme disease-ridden tick when I first learned about the manifesting power of our words—the Law of Attraction ("LOA" for those in the know). The LOA tells me that whatever we give most of our energy and attention to comes right back to us, like some cosmic boomerang. Want more love? Put more love out into the world! Want more compassion? Exude compassion for others!

Want more hate? No freaking thank you. For this people pleaser, attracting any sort of hate, disapproval of any kind, must be shunned the way a vampire avoids garlic-laden aioli. Hence, I've stricken "hate" from my daily vocabulary to avoid any possible hint that the Universe direct more hate in my direction.

Not that the world isn't giving me plenty to hate on these days. See, *e.g.*: the platform formerly known as Twitter. People promoting idiotic pseudo-science test my commitment to avoiding hate about as often as I breathe.

"People are not affecting climate change. You're going to tell me that back in the Ice Age, how much taxes did people pay, and how many changes did governments make to melt the ice?"[90]
U.S. Representative Marjorie Taylor Greene

Just reading about China's latest spree of opening new coal-fired power plants[91] sends my fingers flying on the keyboard. What madness is this, amping up the burning of fossil fuels when we have just a handful of years to stop it?

You have no idea the discipline required for me to bite back using that dreaded hate word on a ~~daily~~ hourly basis. My knuckles are raw from gnawing them instead.

How did we vent our vitriol before the Internet gave us all of these (anti)social media platforms, anyway? I suppose I would've been one of those rabid pontificators on the stump in the town square, banging pots for attention. That still works, by the way. If you're naked and record it for TikTok.[92]

But that darn LOA is always in the back of my mind, challenging me to find the positive in every situation, the good in everyone. It's a Herculean task, considering how many people are fomenting violence against the scientists

---

[90] For the record: No humans were paying taxes in the last Ice Age, some 12,000+ years ago.
[91] Since 2022, China has been on pace to open 2 new coal power plants per week, per the Global Energy Monitor. At the same time, China is making substantial investments in renewable energy projects and electric vehicles. Make this make sense.
[92] Rest easy, kids. That's something you will NOT see me doing.

for employing logic and reason, against the government officials acting to protect democracy, and against journalists daring to report facts.[93]

I admit that the deniers and denigrators, the ones causing us to waste precious time, make me as angry as an airline passenger whose flight was just canceled on Thanksgiving Eve. But—this entire book notwithstanding—I'm more of a lover than a fighter at heart. I pray daily that those who have drunk the Kool-Aid, so to speak, find their way back to the truth, where the Earth is a globe, the moon landing happened and Comet Ping Pong is a pizza parlor, not a child-trafficking operation. Hate and fear mongers who dupe people into disbelieving such basic truths light up my madness meter to the Frenzied Flame[94] level.

I'm noticing that the anger side is winning out over all my loving intentions all too often these days, that Hate word ever lurking on the tip of my tongue. When I caught myself texting HATE HATE HATE to myself, in a failed attempt to get it out of my system, I decided to amp up my Positivity Practice. Take some cues from the Abraham Hicks folks and their advice that we "milk"[95] the good feelings as much as we can, to manifest more and more good. Because I sure do need a good dose of GOOD right

---

[93] And not the alternative kind. When Kellyanne Conway claimed that then-White House Press Secretary Sean Spicer gave "alternative facts," journalist Chuck Todd countered, "Look, alternative facts are not facts. They're falsehoods." That's *journalism*, ladies and gentlemen.
[94] Yes, that's a nod to the game Elden Ring by Bandai Namco Entertainment.
[95] If you're asking if it's dairy or almond milk because you're vegan or if you're incensed because nut-based milks aren't really "milk," I'm sorry I brought it up. We already have enough to be mad about.

now. Like Ikea shoppers need hex keys.[96]

Here goes. Isn't it lovely how well the room-darkening shades keep out the light of day when I'm avoiding it! How wonderfully comfy is this waterbed, even as I struggle to raise myself from it! How thoughtful of the cat to leave a bloody, dead mouse by my bedside!

Not a stellar start, but I must begin somewhere.

I was giving it my all until I caught myself screeching the dreaded H word at my least favorite cat. I hang my head in animal-lover shame, although I do have my excuses—my excuse-making skills merit a gold medal.

See, I discovered that ungrateful cat, the one I rescued from a life of misery at the city shelter, licking my grass-fed, raw milk, organic cheddar. Perhaps this sounds trivial and, in comparison to the ongoing destruction of our planet, it absolutely is. Perhaps you are not acquainted with my fondness for cheese.[97]

Cheese is my love language. It's been there for me like no other foodstuff has. It's delicious for breakfast, lunch, dinner and eating my feelings. Cheeses holds no grudges and never asks, "*Are you OKAY?*" in that way that says, "You look like death's head on a mop stick."[98] I savor and

---

[96] Also called an Allen wrench, but most appropriately named "the devil's crank" by my astute friend Sam.

[97] Yes, the cow milk kind—despite all my eco-despair over all the methane gas that the billion-plus bovines of the world belch into our already greenhouse gas-laden atmosphere. In my eco-defense, I do my best to offset my consumption by paying extra for cheese produced by happy, grass-fed cows.

Please do not waste your breath spouting the virtues of vegan cheese. I have tried so many kinds that I can walk by the vegan cheese display and disparage each by name. None of them lives up to my sharp cheddar or smoked Gouda.

[98] A slick burn from the 1811 Dictionary of the Vulgar Tongue by Francis Grose.

give thanks for every sumptuous bite. Perhaps now you are in a better position to understand my venting on a poor unsuspecting house cat: I HATE YOU.[99]

Oh, I don't really hate her; I hate her *actions*, blah spiritual blah blah.

Still, I surprised myself by using the very "H" word that I claim to shun at all costs. In that moment, I realized that my avoidance of it does not diminish my inner stockpile of vitriol. Just because I don't say it, doesn't mean I don't feel a burning hatred every time I hear about another shooting by some terrorized person fed a steady diet of fearmongering on what passes for news programs these days.[100] Does the world today not provide enough real crises to deplore without so-called news organizations stoking nonexistent flames?

But I digress. Where was I?

Saying only nice words doesn't mean I'm not smoldering in the negative feels. I'm hating on a lot of what I see happening in the world right now. I have done enough therapy to know that burying things tends to amplify them, not make them go away. Kind of like how I used to think wine made all the stress disappear, only to discover a dumpster fire of it in the morning.

---

[99] Not that she cared. When I snatched the plate from her paws, she merely ambled closer, to be in a better position to resume licking. She's as remorseless as a former U.S. President.

[100] I don't know where Kevin Monahan got his news before he shot a young woman for turning into his driveway in Hebron, New York. But I do know a 65-year-old man has been sentenced to serve 25 years in jail because, according to his neighbor, he'd become more and more agitated at people making a wrong turn into his drive. How did feeling "agitated" turn into fearing for his life so much that he took another's? Things I wonder about.

So, I decided to make a list, because that's what I do in times of extreme anxiety; it's one of my favorite coping measures. Making a list gives me the sense of having some sort of control, even when it's a figment of my imagination.[101] When dirty dishes are stacked up in the kitchen sink so high I can't find the faucets, I jot "wash dishes" down on my list. Bish bash bosh, it's already done. In my mind, anyway....

The Hate List.

Just writing those words feels so freeing. Finally! Freed to go about their business, the Inner Cynic and the Constant Critic are pouring margaritas and salsa dancing.

I HATE . . .

1. When I order the perfect latte EXTRA HOT, and it's served lukewarm.
2. When the Wi-Fi goes out in the middle of streaming *Ted Lasso*.[102]
3. When dog walkers put baggies of dog poop in my recycling bin. *EW*.[103]
4. Paying for a streaming service I'm not using but forgot to cancel.
5. My crappy dishwasher that works until it doesn't, so I call the repair person and then it works fine until they leave.

---

[101] What is a figment, anyway? And can you ever have a figment of anything besides imagination?
[102] *Ted Lasso* is essential therapy for me nowadays.
[103] Humans have to sort by hand all of the...*well*...crap that shows up at the sorting centers, people!

6. Phone books. Does anybody even get those anymore?[104]

   So, it starts slowly. I'm still wary of that blowback factor, but I forge on.
7. Plastic things wrapped in plastic which is encased in plastic and packed in plastic.
8. Manufacturers who spew all kinds of plastic crap out into the world without any plan for cleaning up the waste.
9. Supreme Court Justice candidates who claimed Roe v. Wade was "settled law," then rushed to overturn it.
10. Banning a book of literature based on a single, baseless complaint by one whiny busybody.

    Okay, I realize I am veering off topic, unless you consider MAD the topic, in which case this list would never end. I'll do my best to rein in my fuming mind, for your sake.
11. People who harass, stalk, and threaten scientists dedicated to studying the climate, our interconnected biosphere, and life-saving vaccines.
12. So-called news channels that intentionally sow seeds of discord and spew more disinformation than truth.[105]

---

[104] I may have written ad nauseum about my rage for phone books in the "Splayed Out on the Sidewalk" chapter of *Love Earth Now*. Grateful they seem to be a relic of our polluting past now.

[105] Not an insubstantial number, by the way. "The number of partisan-backed outlets designed to look like impartial news outlets has officially surpassed the number of real, local daily newspapers in the U.S." See Axios in References.

THE HATE LIST

13. Oil company executives who knew the damaging effects of climate change way back in the 1970s, but funded decades-long disinformation campaigns denying the science, anyway.
14. CEOs claiming we can't afford environmental regulations while they collect salaries and other benefits worth at least 300 times what their workers earn.[106]
15. Rich people who build extravagant survival shelters instead of funding resilience and adaptation projects that would benefit everyone.

I think I'm getting the hang of this. With all due respect to the Abraham Hicks people, not to mention my own karma, this exercise feels good.[107] Cathartic. As freeing as taking off a double layer of Spanx.

But I am starting to worry that I won't find a place to stop. I can't even type fast enough to record it all, so I step away from the keyboard before the sparks ignite. This laptop is old, but I don't allow myself a new one because Planned Obsolescence would be the next entry on my list.

It's a good thing I meditate. That I vent by writing and doing yoga. That I have an Emergency Coping Strategies List—*of course I have a list*—that gives me multiple exit ramps out of this hate-filled swamp.

Because I don't even want to think about what I'd be doing with all this anger if I'd been taught that guns can

---

[106] The Economic Policy Institute reported that CEOs made 344 times what workers earned in 2022. In 1965 that number was 21 times.
[107] And I use that word loosely. I am not someone who has experienced any kind of "high" from the more strenuous *varieties* of exercise, but I did read about it in a book once.

solve all my problems. If my parents had given me assault rifles instead of Transcendental Meditation classes, I might be in jail right now.

Which is horrifying because I'm so ill-suited for Life on the Inside. How could I survive without my special microfiber pillow, my organic grass-fed cheddar, my introvert Need-to-Be-Alone time? All that and more keeps me on this side of law-abiding. Most of the time. Unless it's a stupid stop sign in a stupid place, then all bets are off.

Ooops, I shouldn't be saying "stupid" because I'm already seeing way too much of that in the world, too.

In any case, I am savoring the sense of sweet release of all my stockpiled anger, much akin to the joy that uber-organizer Marie Kondo promises that a complete purge of my closet would bring. If I could ever accomplish it.

I am ready to be done with the hate-ranting. Pick up that Emergency Coping Strategies List and start checking them off one-by-one, and not stopping, like when I open family-size bag of Doritos, until I feel some peace.

Until I'm on hold for forty-nine hours waiting for a rep at some asshat company to tell me I just missed the window of time to ask for a refund. I really hate it when that happens.

"Kind of a bummer to have been born at the very end of the Fuck Around century just to live out the rest of my life in the Find Out century."
@Merman_Melville on Twitter (now "X") 22 Feb 21

TWELVE
Playlist Track 5: Freshwater

♫ **Recommended Musical Pairing** ♫
"Be the Rain"
~Neil Young and Crazy Horse

Our next already-exceeded boundary in the "Earth Out of Bounds" report concerns fresh water. Humans, they say, can only survive about three days without drinking fresh water. Which makes me thirsty just thinking about it. I'm super suggestible that way.

While the vast oceans on our "blue marble" hold about 321 million cubic miles[108] of water, only 3% of that water is "fresh." Of that tiny portion, 68% of fresh water is locked up in ice and glaciers,[109] and another 30% is groundwater, not all of which is readily accessible. How

---

[108] 1.34 billion cubic kilometers. I have no idea just how large either of those numbers is, but I suspect they are a lot bigger than SCADS.
[109] Which are melting, sending some of our precious freshwater into our salty oceans.

much fresh water in lakes and rivers? It amounts to about a quarter of 1% of all water on Earth. And yet we're depleting those precious fresh water supplies at an alarming rate.

Given its essential nature and increasing scarcity, you'd think we would be treating our freshwater sources with the kind of care we give to the air we breathe, which is also essential for life. Ooops, bad example. Let's move on.

Nope, not in the Backwards World in which we live. We're selling off our fresh water like broken toys at a yard sale. CHEAP. The average cost of tap water in the U.S. is a penny per five gallons, making it 8.4 cents per 42 gallons—the size of a barrel of petroleum. EIGHT CENTS A BARREL.

The cost of oil varies widely, of course, but has averaged about $72 per barrel over the last five years. Eight cents for a life-essential liquid[110] versus $72 for a life-killing substance. No wonder it's easy for some people to think that using water is no big deal. Like home-grown zucchinis in July,[111] it's so cheap, it's tempting to think of it as having no value at all.

That's good old capitalism, I guess, telling us that something being "cheap" somehow overshadows its life-sustaining properties. It's perfectly fine to overconsume and pollute away because our essential-for-life water sells for pennies on the barrel head.

---

[110] The cost of water is even less where companies require no permit to draw water straight from the river or lake.
[111] That's a nod to Barbara Kingsolver in *Animal, Vegetable, Miracle: A Year of Food Life*: "One day we came home from some errands to find a grocery sack of [zucchini] hanging on our mailbox. The perpetrator, of course, was nowhere in sight."

Since water is cheap, we can just keep pumping it out of the rivers and aquifers, polluting it with fertilizers and pesticides, industrial waste streams, urban runoff, and poor sanitation practices[112] with the speed of the Road Runner escaping the clutches of Wile E. Coyote. Rather like saying, "I can't be out of money when I still have my debit card."

How can anyone who has ever been thirsty think this?? I'm betting the THREE-QUARTERS of the people in the world who are water-insecure don't. It reminds me of a worn-out T-shirt I still have and cannot throw out:

"Only when the last tree has been cut down, the last fish been caught, and the last stream poisoned, will we realize we cannot eat [OR DRINK] money."[113]

So, of course, we should just continue all these polluting practices when global warming causes more surface water to evaporate and more frequent droughts, which mean less recharge for our aquifers. Great plan. It's like somebody said the home team is down by 50 points, so let's give the other team the ball / puck / birdie for the rest of the game.[114]

---

[112] The water treatment facilities in the U.S. are aging, many in need of significant repair. See, e.g., Jackson, Mississippi where needed repairs could cost $2 billion.
[113] The "OR DRINK" addition is mine.
[114] I'm no sports expert, so work with me here. Pretend it made sense in your preferred pastime. Unless your sport is swimming or gymnastics or skiing . . . oh, never mind.

Fracking or "hydraulic fracturing" deserves its own spotlight, being the poster child for decimating our fresh water supplies. Hydraulic fracturing is the process of injecting water, sand, and/or chemicals into a well to break up underground bedrock to free up oil or gas reserves. It consumes scads and scads of water—as much as 40 million gallons per well—often in drought-prone places. A report in 2016 reported that nearly 60% of wells fracked between 2011 and 2016 were in "high water stress areas," and 36% were located where groundwater is being depleted. Depleted. As in nothing left for residents, critters, or any other user. As the indigenous Lakota people say, *"Mní wičhóni,"* or "water is life."

All of that overextraction just to mine more fossil fuels, which are then used to power future fracking operations because—guess what—fracking is also quite energy-intensive. In other words, fracking f*cks us on all the fronts.[115]

Consider Wintergarden, Texas, where groundwater aquifers are already stressed from years of drought coupled with pumping to feed cattle and crops. Fracking operators, by far the biggest users of water around here, currently pay about fifty cents a barrel for fresh water. FIFTY CENTS. And there are no restrictions on fracking operations in times of drought, while residents must comply with consumption restrictions.[116] If the residents had to buy bottled water because the tap is restricted,

---

[115] Especially when you consider that fracking is used more and more to source the ethane necessary to produce more non-recyclable, non-compostable, forever trash plastics.
[116] Even in cases of "extreme" drought, fracking is "discouraged" but not prohibited.

they would pay a whopping $504 for a 42-gallon barrel of it.[117] Sure, the bottled water is purified, but is the barrel of water really worth so little when residents don't have enough?

When it's not being depleted by overmining, groundwaters may be contaminated by drilling and fracking operations, leaks from old gasoline storage tanks and landfills, and poor sanitation practices.[118]

Agriculture also places immense stress on our fresh water supplies, consuming 70% of the world's withdrawals—more than any other sector.[119]

Compounding the supply issues, municipal fresh water may be contaminated by leaching lead pipes and failing old municipal treatment systems.

The "Earth Out of Bounds" report says we've already crossed the safe boundary for protecting our fresh water supplies. So why, pray tell, are we still contaminating and consuming them like they are bottomless mimosas at an all-you-can-eat brunch?

I live in Southern California, so I am well practiced in honoring water restrictions in time of drought. I just don't get why we, the burgeoning population of a semi-arid city, aren't subject to even greater restrictions ALL THE TIME. I know, they are unpopular, judging from the

---

[117] Here's the math: a 16-ounce bottle of water currently goes for about $1.50. There's 8 of those in a gallon so that water costs $12 per gallon. $12 times 42 gallons equals. . . $504.
[118] A civil engineer friend tells me that it's common in some European countries to require wastewater to be disposed UPSTREAM of the municipal potable water spigots. What a great ultimate incentive to clean up their wastewater because they will soon be drinking it.
[119] UNESCO, UN World Water Development Report 2024: Water for Prosperity and Peace (Paris: UNESCO, 2024).

derogatory comments from people acting like watering their turf lawn is an enumerated Constitutional right. But is a perfect appearance more important than ensuring we have enough clean fresh water for everyone?

Rhetorical question. I sit in the backyard of Hollywood as I write this, after all.

"In the West, it is said, water flows uphill toward money. And it literally does, as it leaps three thousand feet across the Tehachapi Mountains in gigantic siphons to slake the thirst of Los Angeles, as it is shoved a thousand feet out of Colorado River canyons to water Phoenix and Palm Springs and the irrigated lands around them."
Marc Reisner, author of *Cadillac Desert: The American West and Its Disappearing Water*

So what song am I singing here?

We take our "hazardous household wastes" to a collection center, instead of sending them off to the leaky landfill. Let the hazmat experts decide how to safely dispose of or even reclaim our unwanted paints, thinners, caustic cleaners, batteries, e-wastes, and all the various chemicals we use in our cars, homes, and lawns.

River and lake cleanup days are another great way to protect our fresh water from all the garbage that city living inspires.

All this talk about fresh water has left me feeling

parched. Consider the good works of these group while I go drink a gallon of the stuff. While I still can.

MINNESOTA WATER STEWARDS

The people in the land of 10,000 lakes must know a thing or two about fresh water. The Minnesota Water Stewards program, an initiative of the Freshwater nonprofit, trains volunteers to bring up-to-date knowledge about water issues to city and local government bodies, to create waterwise art projects, to install water-conservation features such as rain gardens and permeable driveways, and to restore prairies. Minnesota Water Steward projects have prevented millions of gallons of polluted runoff waters from entering the state's many lakes, rivers and streams.

Their certification program includes extensive education on topics ranging from hydrology to community engagement.

https://minnesotawaterstewards.org/

THE WATERKEEPER ALLIANCE

"We are a global movement united for clean, healthy, and abundant water for all people and the planet."

Waterkeeper Alliance Vision Statement

The Waterkeeper Alliance is an umbrella organization supporting the more than 300 Waterkeeper groups across the globe. Collectively, they act to protect more than 2.5 million square miles of rivers, lakes and coastal waterways on six continents. Waterkeeper groups organize community engagement and education events, monitor local water quality, advocate for more sustainable practices and conduct habitat restoration projects.

Check out their Take Action page to find their current campaigns and calls for support. Or listen to their "Equity in Every Drop" to discover more ways to advocate for clean water.

https://waterkeeper.org/about-us/

THIRTEEN
My Unbucket List

Yet another #BucketList post flashes across my screen. Yet another glossy image of pink-painted toenails on white sands set against turquoise waters. Another selfie featuring a flushed, beaming face posed in front of the Taj Mahal.

Once upon a time, I'd have been wistful, itching to get out there and travel to the four corners of the Earth. Check boxes off my lists of states, countries, national parks, museums and Japanese toilets[120] I'd visited. I once planned my entire year around my frequent flier mile budget—before parenthood, that is.

Surely a person like me would have a Bucket List full of exotic destinations. I'm a hodophile[121] after all, and a compulsive list maker. I'm hard pressed to make it through the day without making some kind of list, even if it's just a "Things I Forgot to Do Again Today" scrawl in my Bullet Journal. I'm of That Age when the end of life is closer than the start, and a cancer "survivor"[122] to boot.

---

[120] The movie *Perfect Days* got me hooked on the variety and ingenuity of Japanese toilets.
[121] My new favorite word which means "loves to travel."
[122] "No Evidence of Disease," anyway. No doctor says "cured."

Me having a well-polished and researched Bucket List full of exotic destinations seems as predictable as a certain former president denying election results. Clock is ticking, Leutjen.

But a funny, as in wildly unexpected, thing happened along the journey of all my days. As I sat one day to consider what would be on my Bucket List, I found myself writing nothing. I have SCADS of ideas for things certain other people should do, of course, but I found nothing that I needed to do or see for my life to feel complete. I've enjoyed more travel, more adventure, more wild-hare escapades than my little kid self, holed up in her Gladstone, Missouri bedroom, would ever have imagined. Like the kid holding an ice cream cone and demanding another, it seemed a bit unappreciative to insist on more.

Or maybe it's just that I'm finally realizing that, as much I love exploring, I don't require a multi-hour flight to consider my life complete. Plenty of mysteries to solve and adventures to undertake wherever I am. The more I use technology, the more the mysteries pile up.

No one is more surprised to hear me say that than me.

Sure, there are places I would like to see. Alaska remains the only state I've never visited, and its natural beauty is an obvious draw for somebody like me.[123] I never did get to do that writing retreat in Venice, Italy, where I saw myself sipping tiny cups of espresso while the wisdom of Dante poured onto my page. And I have

---

[123] I was ready to hop on the first flight out, back when the comedy-drama *Northern Exposure* series first aired in the '90s—until I learned it was filmed in Washington State.

dreamed of a romantic getaway with my husband, snaking up the steep hills of the Santorini Island caldera, since I first visited in the late 1980s.

I'm also ever more aware of the impacts of our travel, from airplane emissions to pricing natives out of their homes in popular destinations. Case in point: Venice, Italy, where so many locals have been priced out, thanks to foreign property owners and short-term rentals. Much as I love sitting on a white sand beach while sipping a cold drink, I'm starting to think that the best way I can help preserve the remaining pristine places in this world is to NOT GO.

Even a well-meaning ecotourism trip has negative impacts, potentially exploiting local workers, creating trash, and using up precious resources. Eco-resorts built in pristine and fragile habitats open them up to further intrusions. Construction in such areas means hauling in heavy equipment, emitting pollutants by fossil-fuel powered machinery, and building roads into places that may be better off left untouched. A cursory look around the Caribbean reveals far too many failed resorts left to rot. Just like us colonizers to leave our trash behind.

"Ecotourism often serves as the tip of the spear that opens up protected areas for more high impact types of tourism."
Jake Kheel,

Vice Presidente Grupo Puntacana Foundation
Staying closer to home, I spare the locals from the

pollutants generated by my travel, ripping off their resources for my comfort, and relieving them of having to give one whit about yet another white memsahib.

What would be at the very top of my Bucket List is this: that the beautiful, pristine and wondrous places in this world continue to be beautiful, pristine and wondrous when I depart it. If I want butter pecan ice cream tonight after dinner, I gotta take it home and put it into the freezer, instead of snarfing it in the grocery store parking lot.[124]

So I grabbed my trusty notebook and began to scribble.

MY UNBUCKET LIST

What I do NOT need to do to consider my life complete:

1. Skydive, hang glide or bungee jump. Given my fear of heights, these are not a sacrifice.
2. Climb Mt. Everest aka Mount Sagarmatha (Tibetan name) or Mount Chomolungma (in Sherpa language). Add "fear of freezing to death" to my list of phobias and you'll understand why this is also not a loss.
3. Meditate in an ashram in India. Tbh, this one's here because it makes me sound spiritual, but lacking sufficient creature comforts.
4. Go to Burning Man. Too hot for this menopausal.

---

[124] That's a hazard/benefit of carrying my bamboo utensils with me everywhere I go. I'm always prepared to chow down.

5. Swim with the sharks. I've been a lawyer, haha. Some might say I've already crossed this one off.
6. Hike Machu Picchu. More tempting.
7. Explore the Galapagos. So so tempting
8. Visit every U.S. National Park. As a park supporter and a "list checker," this one's hardest to let go.

I'm starting to get into the realm of experiences I'd seriously consider—I'm much more likely to visit national parks than a crowded festival with clogged Porta Potties—and still, I'm at peace with my Unbucketing. A friend just shared stories from a recent trip to the Galapagos, and my younger self would have taken copious notes, penciled it onto her someday calendar. Lovely as it sounds, I still mean to find peace with being Where I Am. Because my children's future may depend on it—both my being Present, instead of mindlessly checking out, and also by not harming the people and places that make this Earth the treasure that it is.

Here's my complete Bucket List then.

BUCKET LIST

1. Do all I can to ensure that the enchanting places and one-of-a-kind opportunities that have been available to me are still here for my children and their children whenever they contemplate their own version of a bucket list.

It's not a satisfying list, given my love of long and laborious itemizations, made in contrasting colors and scripted fonts. But it's the box that I most want to check off at the end of this life.

As I am contemplating the end to this chapter, Mary Shaugnessey begins telling me her story via the Moth Radio Hour podcast on NPR. In 2020, Mary was diagnosed with thyroid cancer, stage 4, prognosis: 4 to 6 months to live. From the depths of despair, she summoned the will to raise her hand. She called upon her "people" to send her messages of love and healing. The outpouring of support transformed her, and she learned to celebrate a life that brought her so many loving relationships.

Mary went on to say that she didn't have a Bucket List, either. Why? "Because, in the end, I don't think that life is measured by how many big adventures we squeeze in... If the object of this crazy game we call Life is to give and receive love, I can tell you with one hundred per cent certainty, I am WINNING." The crowd roars. Standing ovation. Winner.

The host proceeds to tell me that... Mary passed on three months after recording this story.

Did I have to swerve my car to the side of the street to sob? You bet I did. I'm 17 months out from the end of my cancer treatments. I've already been given so much more time than Mary that I'm feeling guilty—she was just 57 and her children are much younger than mine.

Even more than teaching me to be grateful, Mary validated my sense that "winning" this life, feeling good about what I've done with my one wild and precious life, doesn't depend on the "big adventures." My concise Bucket List stands.

Not that I'll ever be content to while away my days in an armchair. I'm too antsy. I still love exploring, and I aim to go someplace new often—both for the sense of adventure but also because it's good for cognitive health.[125] Living in my hilly corner of Los Angeles, I don't have to go far to find myself in a foreign land (as in, I'd be lost without Google Maps). I discover new vistas within a couple of miles whenever I go out to make a simple Buy Nothing[126] pickup—and I've lived in this area for twenty-five years.[127]

Keeping my adventures closer to home not only lessens my carbon footprint but forges deeper connections in my local geography. Driving around my local community introduces me to new neighbors. Hiking in the local hills educates me about the local flora and fauna that warrant my concern and protection. Even a slog down my street informs me about growing seasons, as neighbors set out boxes of excess pomegranates, zucchini, or lemons from their gardens.[128]

All that said... I do have serious reservations about everything I've just said in this chapter. I know how contrary the Universe can be. As soon as I say, "I'm content to stay home," then along come a million reasons

---

[125] "They" say it's good for your brain to take different routes, but I suspect wandering was a more effective therapy back before Google Maps. I mean, how challenging is following step by step instructions? But then Googlia does have her moments. I've ended up in some interesting places that she assured me were on my way home. Ha.
[126] What's Buy Nothing? See the "Playlist Bonus Track" chapter.
[127] It really is that crazy, trying to pass cars on narrow, two-way streets barely wide enough for a single car that suddenly become dirt trails.
[128] God bless them! My gardening skills leave everything to be desired.

why I must be on the road the next 300 days out of 365.

So, please don't @ me if you see me hiking a faraway ruin someday. Because the Universe has its own mind, and who am I to question the Divine Will?

And then this temptation just landed. I sure would love to sample organic chocolate at my pal Gary's eco-farm, Green Acres Chocolate Farm and Nature Preserve, on the banks of Dolphin Bay in Bocas del Toro in Panama someday. That would totally be on my Bucket List if I could ever bring myself to pen a #2.

"After the midterm elections, my advisers asked me, 'Mr. President, do you have a bucket list?' And I said, 'Well, I have something that rhymes with bucket list.'"

U.S. President Barack Obama[129]

ECOTOURISM DONE RIGHT

If you've got a Bucket List full of travel destinations, you'll get no judgment from me, whatsoever. I've enjoyed the traipsing I've already done, from the Philippines to Paris. Judging someone else's travel choices would be rather hypocritical, wouldn't it[130]

---

[129] I'm with Obama. I definitely have a list that rhymes with Bucket List.
[130] There's SCADS of that happening already.

Ecotourism, done right, can be beneficial for local residents and habitats alike. It educates travelers about the beauty, the wildness, the species, and the habitats that warrant our care and concern. It often raises funds for conservation initiatives.

But how does anyone know which resorts and tours are truly sustainable, respectful of the people and critters who live there?

1. **Certifications**. Look for destinations that have eco-certifications or are recognized by reputable organizations.

2. **Local Involvement**. Involvement by the local community is a sign that locals are being treated fairly.

3. **Environmental Practices**. Investigate the destination's environmental policies. Are they minimizing waste, conserving energy, and protecting natural habitats?

4. **Size**. Consider the maximum number of visitors the destination can handle without harming the ecosystem or local culture. Sustainable limits are crucial.

5. **Reviews**. Check for reviews before choosing a destination or guide.

Gary Mitchell, President of that Green Acres Chocolate Farm I mentioned above, recommends asking how the eco-resort gets or produces power. Many run on diesel-powered generators that not only create harmful emissions but require delivery of fuel, often by greenhouse-emitting vehicles. He also recommends investigating the food served to guests. While lavish meals may be enticing, they put a strain on local resources and also require more supplies to be delivered by trucks or boats. Resorts serving local, vegetarian food create a much lower impact.

For more resources, see: "Travel Responsibly: A Guide for Sustainable Hotels and Tours" by Susan Portnoy in *The Insatiable Traveler*. https://theinsatiabletraveler.com/travel-responsibly-guide-for-sustainable-hotels-tours

GREEN ACRES CHOCOLATE FARM

Green Acres Farm, nestled alongside scenic Dolphin Bay in Bocas del Toro in Panama, checks off many of those boxes. The facilities are powered by solar panels, collected rainfall supplies all the water they use, and they work closely with local indigenous peoples.

But Green Acres isn't just another eco-resort. Gary Mitchell and his devoted team work so tirelessly for conservation that I wonder how they have the bandwidth to host guests. They donate to local communities saplings from their native almendro trees (*Dipteryx panamensis*), a critical species with medicinal properties that was almost harvested to extinction, thanks to its termite-resistant wood. The trees also provide critical habitat for the endangered Great Green Macaw, which Green Acres also supports by fundraising to build a breeding center. Those are just a few of their current eco-projects, along with producing organic chocolate from three types of cacao, including the rare and highly prized Criollo strain.

They also treat their guests well, maintaining a five-star rating and the coveted "2024 Travelers Choice Award" from Trip Advisor. The two-hour tour of their 30-acre farm includes strolls through lush tropical forests and botanical gardens full of brilliantly hued exotic flora, plus wildlife encounters with howler monkeys, sloths, poison dart frogs, toucans, and a variety of other birds. Guests may may also visit indigenous villages and sample the award-winning chocolate. If you decide to stay in the off-grid Rainforest Cabin, you'll enjoy sumptuous vegan meals and Gary's impeccable hospitality.

Yeah, that's where I'd go if I had a real bucket list. If you go, please give Gary a hug from me.

https://greenacreschocolatefarm.com/

FOURTEEN
Playlist Track 6: Climate Change

♫ **Recommended Musical Pairing** ♫
"Our Planet"
~Chris Webby (feat. Bria Lee)

We've come to the last of the six boundaries in the "Earth Out of Bounds" report that have already been exceeded. Did I save this final out-of-boundary measure, the poster child for Woke-ism, for last on purpose? You bet I did. Because I am weary[131] of people saying they don't "believe" in climate change, as if that's an excuse for doing nothing to care for the environment that sustains Life as We Know It. If 80+ years of climate science,[132] the six detailed climate assessments from the

---

[131] "Weary" being code for fed-effing-up.
[132] In 1938, British engineer Guy Callendar reported increases in both global temperatures and atmospheric CO2 concentrations. He smelled a rat. See "BBC Climate Change Timeline" in Resources.

Intergovernmental Panel on Climate Change,[133] and SCADS of peer-reviewed studies aren't convincing, how about doing something about biodiversity or freshwater supplies or any of the other dire crises coming at us like a runaway bullet train?

Because, again, for those in the back, I'm not here to convince anybody that the planet is warming or that humans, with our fossil fuel-burning ways, are the cause. If you haven't read, learned or experienced anything that has convinced you of this basic and disturbing reality, nothing I say here will change that. Skipping ahead to the next chapter is totally allowed.

For those still reading, let's unpack this global temperature thing. Scientists say we've already raised the average global temperature one degree Celsius (that's 1.7º Fahrenheit). Many scientists say we need to limit the increasing average global temp to a maximum of 1.5ºC (2.7ºF) if we want to keep the impacts at a painful-but-survivable level.

But unless we change things—FAST—our great-grandchildren could be facing even more dire consequences with an increase of 2.0ºC, 3.0ºC or even 4.0ºC (7.2ºF). Which would be survivable in the way that people endured the desert wastelands in the *Mad Max* movies.

Hundreds of millions of lives could be lost due to sea

---

[133] The IPCC includes representatives from 195 member countries. If these reports are all part of a giant conspiracy by the Woke, then I want to know who is coordinating the effort. Getting two people to agree on anything is near impossible. Who is the mastermind who could get all these reps from around the world to agree on a single fake news narrative about an issue as big as this one?

level rise, killer heat waves, and the inability to grow food in blistering heat, drought and deluge. If you think the southern border of the U.S. is busy now, just imagine it when the homelands of millions become too hot or the rainfall too erratic to grow food. The World Bank estimates that 17 million Latin Americans could be forced to relocate between now and 2050, due to reduced crop productivity, rising sea levels and storm surges.

"Consider that for generations your family has been growing certain crops and has learned the rhythm of the seasons and has a system for how and when to harvest. And suddenly those seasons are changing. And it's not just one season shifting and then going back to normal, but shifting permanently and continuously. It's very difficult to be able to adapt to that kind of rapid change in a way that in the short term still leaves you able to sustain your livelihoods and feed your family."

Gabriela Nagle Alverio
Duke University J.D.-Ph.D candidate

But how can any of that be? If the temperature inside my home varied by a degree or two, I wouldn't even notice. At any given moment the temperatures around the globe could vary by a hundred degrees Fahrenheit. How can an increase of a degree or two in the *average* temperature wreak such havoc?

I suppose it's a matter of the sheer magnitude of measuring anything across the space of an entire Earth.[134] I'm no statistician, but I think of it like data that says the "average" salary at a company is $196k, but the hundred employees make $50k a year and the CEO makes $5,000,000. Hey, boss, can I have that "average" salary, please?

It's also true that those averages don't reflect conditions at any one point. The Arctic is heating at a rate FOUR times faster than the rest of the planet, which terrifies me more than my freezer going out on the hottest day of the year.

It's still hard to fathom how such a small change can make such extreme results. Then again, isn't it extraordinary what changes we're experiencing from the small increase we've already triggered? Every year end brings another announcement it was the "hottest year on record." The last "cooler-than-average" year was 1976. Millennials, Gen Z, and Gen Alpha have never experienced such a thing.

My college-aged self, the one who studied climate change back in the early 1980s, is apoplectic, demanding to know WTF happened? Why didn't we, me included, do more back then, instead of standing back and watching this, the deadliest Mousetrap[135] game unfold?

Let's review where we are. Because, if you're like me,

---

[134] For a better explanation, see: Climate Change: Global Temperature | NOAA Climate.gov

[135] The Mousetrap game of my youth involved a sort of Rube-Goldberg contraption, which had to be assembled in a specified order to play the game. In other words, we could never play it because inevitably a piece was missing.

you've been too busy changing passwords to take it all in. Or maybe that's just my excuse. Delving into this depressing research, digging into all the facts and figures proving what we should have known all along, sickens me more than cleaning up projectile vomit. So, I've armed myself with a yard-long baguette to gnaw,[136] and steeled myself for this awful journey. I suggest you fortify yourself with whatever calms you before plowing through this chapter.

WHERE WE ARE.

Melting ice. We've already caused the melting of enough glacial ice from Antarctica and Greenland—some 5,000 gigatons—to fill another Lake Michigan in just the past 16 years. Continued melting in the Arctic will release a whole lotta mercury that's been sequestered in the permafrost for millennia. As ice melts into the rivers, all that mercury becomes a health hazard for the 5 million people living in the Arctic zone. And a food hazard for all the rest of us, as mercury accumulates in the fish we eat. Bon appétit.

Heat waves. Heat-related deaths are going up, up, up, too. Some twenty-three hundred people died in the U.S. of heat-related causes in 2023, the highest number in the last forty-five years of recordkeeping.

Superstorms. Superstorms keep supersizing. Warmer ocean waters make for more intense hurricanes, and guess what's making warmer ocean temps 400 times more likely?

---

[136] I find gnawing an effective means of diffusing fury. Or maybe it's just an excuse to eat carbs.

Case in point: Hurricane Beryl intensified with breakneck speed in June-July 2024, forming over record-breaking warm ocean waters. With wind speeds clocked at 165 mph, Beryl was the earliest-in-the-season Category 5 hurricane ever recorded. Beryl caused some sixty deaths and $6 billion in damages in the U.S. And that was just the start to the 2024 hurricane season.

That was the most recent hurricane when I first penned this chapter, but Hurricane Helene of September 2024 necessitated a late update. As of this writing, Helene intensified even more rapidly than Beryl, and the storm surge reached 15 feet in some areas, making it one of the highest in the region, based on records dating back to the 1800s. As of this moment, just days since Helene hit, the death toll from Helene approaches two hundred, and countless folks remain stranded in rural mountainous areas. The images of entire towns submerged and people huddling on rooftops haunt me. Reports of bodies found in trees send me to my closet floor weeping. WHAT HAVE WE DONE?

Early estimates of total damage and economic loss from Hurricane Helene amount to $200 billion—almost none of which will be covered by insurance. From one hurricane.

Meanwhile, meteorologists monitor Hurricane Milton, as it barrels across the Gulf of Mexico toward Florida. Hurricane season doesn't end for two more months.

Floods. Hurricane Helene also caused devastating floods across some 800 miles in Florida, Georgia, North and South Carolina, Tennessee and Virginia, the full extent of which is not known as of this writing. Severe thunderstorms in the northeast U.S. in August 2024 delivered a "1,000-year" rainfall, resulting in flash flooding and multiple water rescues. Many areas were already "overwatered," so to speak before Hurricane Helene arrived.

Torrential rains in Iowa, South Dakota, Nebraska, and Minnesota in June 2024, resulted in road and business closures, the collapse of a train bridge, the evacuation of hospitals and nursing homes—and nearly caused a dam to collapse. That's just in the U.S.

Repeated catastrophic downpours in the Guangdong and Hunan Provinces in China caused so much flooding and mudslides that more than 110,000 residents had to relocate. Flooding in Brazil, the worst in 80 years, caused multiple deaths, widespread landslides and a dam collapse—following on the heels of similar calamities in July, September, and November 2023.

Those are just a few ~~highlights~~ "lowlights" in the flooding category from 2024, a year in which, as of this writing, three months remain.

Droughts. Droughts in the western U.S. since 2000 have reached legendary status, each one topping the list of "worst ever." Droughts are drying up drinking water supplies, and decimating water supplies for California's farms which supply half the U.S. with nuts, fruits and vegetables—and some 20% of the nation's milk. Groundwater levels have sunk lower than wells in some places, thanks to over-pumping to compensate for long years of drought.

But droughts aren't confined to the Western U.S. As I

write this in early September 2024, the U.S. Drought Monitor (USDM) map shows zero areas of drought where I sit in Southern California, thanks to a couple of bizarrely wet winters. But parts of Ohio, West Virginia, Mississippi, Texas, Oklahoma, Tennessee, Kansas, Montana, Wyoming, Oregon, Washington, South Dakota, Idaho, and Hawaii are all experiencing severe to exceptional drought conditions. Ohio and West Virginia face exceptional drought conditions for the first time in the 25-year USDM history. Not the kind of records we want to be setting.

Warmer air worsens the impacts of droughts by causing even more evaporation, desiccating soils, and drying up fresh water sources. It's all bad news for crops, livestock, and wildlife. One hundred per cent of topsoils in West Virginia are currently rated dry or very dry, making for widespread crop losses. Drought impacts in Oklahoma, Tennessee and northern Mississippi mean little to no vegetation for livestock, widespread crop losses, dried-up farm ponds and increased risks of wildfires.

Wildfires. As a fervent forest lover, this one guts me. I may need a fermented beverage to go with my baguette.

The number of wildfires in the western U.S. has doubled between 1984 and 2015 thanks in large part to climate change. Warmer temperatures dry out organic matter in the forest so it burns more readily. They also increase the length of the fire season. All of the top ten years with the most acreage burned have occurred since 2004—coinciding with many of the warmest years on record. Eighteen wildfires in the U.S. since the year 2000 caused more than $1 billion in damage. EACH.

Earlier this month, an inferno raged in a mountainous region about an hour from my house. I rely on those mountains for my escapes-for-sanity, my family camping trips, and writing retreats. The ski resort we visited just months ago escaped destruction by shooting snow cannons at the fire. That's just the closest blaze. In total, three major conflagrations ravaged some 100,000 acres of precious Southern California forestland. In one month. Meanwhile, Hurricane Helene was doing her worst on the East Coast. It's been a helluva month.

Wildfires are also becoming more common in the Arctic, which is warming faster than the rest of the world. Let that sink in. Wildfires in the ARCTIC. That warming thaws the permafrost and exposes organic material which dries out and becomes fuel more for more wildfires. All of which releases a lot of previously stored carbon, exacerbating climate change and perpetuating the whole disastrous cycle.

We're incinerating trees and forest habitat like they're garbage, instead of the precious resources we need to shade us, to provide critter homes, to prevent soil and water losses, to suck up our oh-so-excess CO_2, to process air pollutants, and reduce our stress. ~~Great~~ Horrific plan.

Sea level rise. Sea levels are rising, thanks both to the melting of a whole lot of ice but also because warmer waters expand. Coastal communities around the world are making their plans to shore up or ship out. The Yup'ik people of Newtok, Alaska surmounted significant logistical, financial, and emotional hurdles to move their community to a new location nine miles away. Their ties to their homeland run generations deep, so this is not your average relo for a new job. They will not be the last people forced to abandon lands they called home for decades or even hundreds of years.

[takes a swig]

That's just a sampling of what humans around the world have *already* experienced, as a result of our fossil fuel additions. If all these extreme weather events make life difficult for people to navigate, imagine the impact on animals. Droughts dry up their (formerly) reliable watering holes, floods and fires destroy their habitats, and heat stresses them in all the ways. Animals that evolved to fill an ecological niche are finding themselves ill-suited to the places they've traditionally called home. Many are moving to new habitats, if they can find any available.

- Unpredictable ice patterns are impairing the ability of some whales to migrate, causing some to become trapped in the ice. The fluctuating extent of ice also impacts the supply of their phytoplankton food sources, resulting in food insecurity and reduced reproduction.
- Forest elephants in Gabon, Africa are going hungry as global warming reduces the yields of the fruits they eat. I can't bear to view the images of their bones poking through their thick hides. The rainforests of Gabon comprise one of the last remaining strongholds for forest elephants. Those fruit-bearing trees that feed elephants depend, in turn, on the animals to spread their seeds. The entire rainforest ecosystems of Gabon are in peril.

- Giraffes may tolerate warmer temperatures better than other species, but heavier rainfall and flooding reduce their numbers, possibly because of increase in parasites and disease. Giraffe populations in the wild have already decreased some 40%, with only some 100,000 individuals remaining in the wild.[137]
- Reduced and unpredictable sea ice makes hunting for their favorite food, seals, difficult for polar bears. If the moms don't have enough food, they don't have enough body fat to produce sufficient milk for their young to survive the long winter months. As a mom...this haunts me.

That's not even a fraction of the affected species, but my gnawable baguette is long gone and my fermented beverage glass licked dry. For a quick sum up, consider this: nearly every one of the 1,497 animals on the U.S. endangered species list faces climate-relate threats. So do another million species around the globe. And if that sounds good to you because we're freeing up space for more mini-malls, I have to wonder why you are still reading this book.

Suffice to say, I suggest we look around and see how the changing climate is already affecting our roommates before we decide it's all a hoax. Or that the "economy" demands we stick to the status quo. Because responding to any one of these events ain't gonna be cheap. Housing

---

[137] Did you know that giraffes have the same number of neck vertebra as humans? Giraffe vertebrae are just a lot bigger.

is already expensive and open space available for animal habitat is already scarce. How much will a new home cost if we must relocate a gazillion people? How many more wars will be fought over the dwindling supply of fresh water, the inability to grow crops in previously fertile fields, and the increased need for medical care and supplies?

That's a lot. I know.

Let's take a collective breath. Maybe time for a walk around the block, a cup of herbal tea or a trip to the rage room. Because that's just the briefest summary of the impacts we've already set in motion. If we don't make big changes, all those impacts will ratchet up exponentially like the price of toilet paper in a pandemic.

It all sounds (and is) dire, but how much of a difference that half or whole degree of warming is is still hard to get my head wrapped around. Here's one thing I know for sure: extreme heat is bad for my temper and I'm already MAD. I suggest you do not come at me with your conspiracy and hoax bollocks when it's hotter than, say, 80 degrees outside. Yes, that's Fahrenheit.

WHERE WE ARE HEADED.

Where could we be headed if we don't curtail global warming? The short answer is to tally up all those existing impacts and ratchet them up by an uncomfortable amount.

Heat waves. Heat waves, for just one example, could last 17 days longer than currently if the average temps top out at 1.5° C. Menopausal women beware. They could last an extra 35 days at 2.0°C. Beware of menopausal women.

Suffice to say, the pileups in the grocery store freezer aisle will be massive, people waving those doors open and closed like three-year-olds who just learned how doors work. That's where I'll be anyway.

Crop losses could skyrocket. Not only will previously fertile fields become uncultivatable, due to floods, droughts, or both, but we may have to say good-bye to some favorites like coffee, grapes (aka wine), and chocolate. Coffee, cocoa plants, and grapes are very sensitive to temperatures—like a lot. Warmer temperatures also encourage pests and diseases that attack grapevines. Do I have your attention now?

"[T]he consequences of climate change on coffee production triggers other risks associated with the soil, water, crop, and nutrient management: drought, salinity, biodiversity decline, suitability losses, change of species seed availability, resistance to abiotic and biotic stressors, etc."

"A Systematic Review on the Impacts of Climate Change on Coffee Agrosystems"

While we in the more developed world might consider these losses tragic, the reduction in maize crop yields would be globally disastrous. It's a crucial staple crop that provides at least 30% of the food calories for more than 4.5 billion people across 94 developing countries. A recent study found that maize production may decrease by 30%, many in Latin America. Studies indicate the yields could decrease by 24% by the end of the century, with significant reductions beginning in 2030. If you think the U.S. southern border is busy now, imagine it when a few billion more people become food insecure.

Human Health Threats. Disease-carrying pests like mosquitos will flourish in the warmer world, and their ranges will expand into new territories. Ticks, mosquitos, and pine beetles are moving northward, into Canada and northern Europe, places which were previously too cold for them to survive. They bring dengue, malaria, Zika, Lyme disease, and more with them.

SUM IT UP

For those who have been skimming through the data-laden paragraphs—and I don't blame you at all—the Australian Climate Council has created an at-a-glance infographic that sums it up.[138]

The difference between 1.5°C warming and 2.0°C warming means:

- Extreme heat events 2.6 times worse.
- Loss of vertebrate species= 2x more
- Loss of plant species = 2x more
- Loss of insect species = 3x more
- Amount of permafrost that will thaw=38% worse

---

[138] If you're ready to throttle me because I didn't tell you about the summary infographic sooner....sorry.

- Reduction in maize harvests in the tropics=2.3x worse
- Further decline in coral reefs = 29% worse
- Decline in marine fisheries=2x worse

AN AWFUL CAVEAT

All those educated guesses about the impacts of further global warming depend on simulation models, which researchers are creating and updating as they collect more data. And I thank them for (what must be) their depressing labors. As of today, however, those models may not be able to account for cascading disasters and feedback loops, where one bad thing triggers another horrible thing. Like when the entire organizational system in my closet collapsed because I hung up one more plastic hanger. In other words, the impacts could be far worse than anticipated.

Here's just one truly horrific example. The Atlantic Meridional Overturning Circulation (AMOC), a critical system of currents in the North Atlantic, could slow or even stop functioning altogether. WTH is the AMOC, did I hear you ask? I didn't know either until I learned that, like reproductive freedom, it would be a terrible thing to lose.

The AMOC circulates warm, salty water from the tropics up to the North Atlantic, where the water cools, becomes denser and sinks. The water then flows back toward the tropics, creating a continuous loop.

Sounds like a nice ride. Why do we care? The AMOC plays a big role in managing the distribution of heat across the planet, as well as supporting scads of marine habitats, among other things beyond my comprehension. Scientists are concerned that all the melting glaciers in the Arctic and Greenland could dump so much freshwater into the salty ocean in the North Atlantic that AMOC weakens significantly, no longer circulating as it currently does. Or it could stop circulating altogether. Which could lead to more frequent and far more disastrous weather events, among other things. It's considered a major "tipping point," a sign that we've really gone too far with our fossil fuel burning habits.

I don't claim to understand it all, but I do quake in my eco-boots to think that we humans—one of 5 billion species that have evolved on Earth over three billion years—could disrupt something as vast as the entire Atlantic Ocean in just the last 250 years.[139] We're not just the bad roommates trashing the apartment, but the heinous kind who burn down the whole city.

So where does all of that leave us? Andrew Boyd, author of *I Want a Better Catastrophe*, summed it up when he said: "We're fucked."

Are we drinking / smoking / gnawing yet?

"The era of global warming has ended; the era of global boiling has arrived."

United Nations Secretary General
António Guterres, July 2023

---

[139] Since the Industrial Revolution.

Okay. That's all horrific, which is as understated as comparing a heart attack to a hangnail. But what about the steps we've already taken? Despite our sluggish responses, countries, companies, and individuals have been making certain strides toward a less fossil-fuel dependent world. What impact have all our paltry-but-well-intentioned efforts already had?

I know the air in Los Angeles still exceeds health standards on the regular,[140] but air pollution here is a fraction of what it was back in the 1970s when schoolchildren weren't allowed outdoors to play. Is it like that, bad but still better?

Sort of. The 2023 State of the Climate Action Report sets out what we must accomplish to prevent the most devastating impacts of climate change, along with a summary of how far we the People of Earth have come in meeting them. Of the 42 "indicators of progress," we're on track to meet only ONE of them—and that is the sale of electric cars. Which is about as encouraging as a kid's report card that says they are failing all subjects except shopping.

---

[140] Here's a surprising fact this former environmental attorney just learned. Our personal hygiene habits are now a significant source of air pollution in today's Los Angeles. Everyday petroleum-based products like deodorants, perfumes, and soaps produce as much volatile organic compound (VOC) pollution as vehicle tailpipes do.

This year's State of Climate Action finds that progress made in closing the global gap in climate action remains woefully inadequate... efforts to end public financing for fossil fuels, dramatically reduce deforestation and expand carbon pricing systems experienced the most significant setbacks to progress in a single year, relative to recent trends."

2023 State of the Climate Action Report

So what song am I singing here?

I'm reading a book called *Sea Change: An Atlas of Islands in a Rising Ocean* that showcases some 49 islands and island nations around the world most impacted by rising sea levels caused by climate change. More than just an atlas of maps, author Christina Gerhardt shares the facts, history, geography, flora and fauna, and lore of each island, giving us an insider view of the people and their culture. Samples of their poetry and stories allow us to hear directly from the hearts of the islanders. This atlas is a testament to the will of the people fighting for their homelands as well as an elegy of what will be lost.

I may not be able to prevent the loss of their lands as the oceans rise to cover them, but I can honor their history, their humanity, their dignity by learning about them—people I might not have known existed otherwise. I've challenged myself to read a chapter a week, and then carry the affected people in my heart until the next chapter. They deserve nothing less than my full attention.

What would you do if people on the other side of the world did something to destroy your home? How many lawsuits would we litigious Americans file? How many college campuses would erupt in protests? How many GoFundMe campaigns would spring up to help the displaced? How many guns would we aim at the culprits?

Rhetorical question.

SHAREHOLDERS ACT

Tackling climate change as an individual feels bewildering. Hopeless. Impossible.

Sure, the basic task sounds simple—just stop burning fossil fuels!—until you consider the deeply-entrenched systems and subsidies that support and protect the vested interests of the ExxonMobils of the world. What can any one of us Davids do to stop the Goliaths? What stones do we have to sling that will knock down the giants pushing their climate-destroying wares?

If you're a shareholder of any of these publicly traded fossil fuel giants, you have a stone in the form of a vote. Some shareholders are organizing and demanding these behemoths acknowledge the risks of climate change and the part their products play in causing it.

Shareholders of BP, ExxonMobil, Occidental Petroleum and the PPL Corporation have all voted in favor of such disclosures in recent years.

What's more, a 2021 study published in *Harvard Business Review* found those climate change-related disclosures resulted in a net positive impact on the firm's stock price. It seems investors value transparency over undisclosed risks. Who knew?

Future legislative actions on climate change may inhibit the sale of fossil fuel assets, thereby diminishing if not eliminating shareholder asset values. As an investor, I'd want to know the financial risk of that, wouldn't you?

Are YOU a shareholder? Check with your financial advisor or do some research on groups that may be active in your area of investing.

SHAREHOLDERS DIVEST

Dumping investments in environment-destroying assets is another way to support action on climate change. Many of the world's largest financial institutions have pledged to do so, and you can, too. [141]

Why let the climate-destroying assholes use YOUR money to fund their polluting ways? See these resources to find environmentally friendly investments. Note that I have not used these services, nor am I endorsing these services. DO YOUR RESEARCH and consult a trusted advisor.

---

[141] Not everyone is living up to their commitments, however. See the 2023 "Who's Managing Your Future" report for an assessment of how some 30 big asset managers continue to invest in new fossil fuel projects, despite their commitments. Looking at you, BlackRock, Vanguard, and State Street, in particular.

#MovetheMoney campaign, sponsored by the "We Don't Have Time" organization (see Playlist Track 9: Ozone chapter) assists shareholders in identifying whether your bank, pension fund or other investments still invest in fossil fuels—or have plans to transition away.

The Green Century Fund is a nonprofit that advises clients on environmentally-sound investments. The group also takes action to persuade companies to adopt more sustainable policies that protect the environment.

https://www.greencentury.com/about-us/

As You Sow: A shareholder advocacy group that also provides ratings of mutual funds for various causes, including support for biodiversity, the circular economy, climate, sustainable farming and more.

Want to find mutual funds that are free of investments in deforestation or fossil fuels[142]? Check out their scorecards.

https://www.asyousow.org/about-us

---

[142] As You Sow also rates funds on issues relating gender equality, gun control, prisons, and tobacco.

Tell Them

I prepared the package
for my friends in the states
First—the dangling earrings, woven
into half-moons, black pearls glinting
like an eye in a storm of tight spirals

Second—the baskets
sturdy, also woven
brown cowry shells shiny
intricate mandalas shaped
by calloused fingers

Inside I write a message:

Wear these earrings
to parties classes and meetings
to the corner store the grocery store
and while riding the bus

Store jewelry, incense, copper coins
and curling letters like this one
in this basket

And when others ask you
where you got this
you tell them

They're from the Marshall Islands

Show them where it is on a map
Tell them we are a proud people
toasted dark brown as the carved ribs
of a tree stump

Tell them we are descendants
of the finest navigators
in the world

Tell them our islands were dropped
from a basket
carried by a giant

Tell them we are the hollow hulls of canoes as fast as the wind
slicing through the sea

We are wood shavings
and drying pandanus leaves and sticky bwiros at kemems

Tell them we are sweet harmonies
of grandmothers mothers aunties sisters—
songs late into night

Tell them we are whispered prayers the breath of God
a crown of fuchsia flowers encircling
Aunty Mary's white sea foam hair

Tell them we are styrofoam cups of koolaid red waiting
patiently for the ilomij

PLAYLIST TRACK 6: CLIMATE CHANGE

We are papaya golden sunsets bleeding into a glittering
open sea
We are skies uncluttered
majestic in their sweeping landscape We are the ocean
terrifying and regal in its power

Tell them we are dusty rubber slippers swiped from
concrete doorsteps
We are the ripped seams and the broken door handles of
taxis

We are sweaty hands shaking
another sweaty hand in heat
Tell them we are days and nights hotter than anything
you can imagine
We are little girls with braids
cartwheeling beneath the rain

We are shards of broken beer bottles
burrowed beneath fine white sand
We are children flinging
like rubber bands
across a road clogged with chugging cars
Tell them
we only have one road

And after all this
tell them about the water—how we have seen it rising
flooding across our cemeteries
gushing over the sea walls
and crashing against our homes

Tell them what it's like
to see the entire ocean level with the land

Tell them
we are afraid

Tell them we don't know
of the politics
or the science
but tell them we see
what is in our own backyard

Tell them that some of us
are old fishermen who believe that God
made us a promise
Tell them some of us
are a little bit more skeptical

But most importantly you tell them
we don't want to leave we've never wanted to leave
and that we
are nothing
without our islands

> "Tell Them" from *Iep Jaltok: Poems from a Marshallese Daughter* © 2017 by Kathy Jetñil-Kijiner. Reprinted by permission of the University of Arizona Press.

FIFTEEN
Activism For Aging Ragers

Stand back, world. The activist elders have the time, the passion, and, yes, the compression socks to do some major rabble rousing. And now they're organized, too. Get between a climate granny and her pocketbook full of grandbaby pics at your peril. A recent study determined the average age of climate activists is 52—and 24% are 69 and older. Yes, that number is SIXTY-NINE.

That statistic made me pause—did it for you, too? Much as I enjoy an active rant, I have reached that age group where any physical activity requires special consideration of my . . . *situation*, shall we say. This comes to mind as I wince from rotator cuff pain, the result of sleeping on my arm wrong one night six months ago. Seriously.

I recall with fondness the old cheerleading days when the worst hamstring pull would heal overnight, instead of crippling me for a month. Good times. These days, I'm doing good to remember to apply that CBD-infused muscle balm before repeatedly and vigorously bashing a pizza box on the recycle bin—a rant inspired by a certain U.S. Presidential debate circa June 2024.

I am in awe of elders who aren't content to sit back in their rocking chairs and witness the ruin of the planet (and certain rights we presumed inalienable). Nope! Many of us older people are taking up the activist mantle. Grannies at women's marches carry signs full of curse words I'm sure my grandmother never uttered. Dudes older than our current President, Joseph R. Biden[143] man the protest lines outside the Chevron plant. The Elders for Climate Action, the Society of Fearless Grandmothers, and the 1000 Grandmothers for Future Generations are not to be trifled with.

> *"When I look my grandchildren and my great-grandchildren, my children, in the eye, I have to be able to say, 'I did everything I could to protect you.'"*
> Hazel Chandler, 78-year-old activist-volunteer in Arizona with Elder Climate Action.

As I sit here icing my aching rotator cuff, I am considering what special adaptations we "aging ragers" should consider before putting our *maturing* bodies on the line.[144] While I've attended several protests, I've never been arrested. So, I interviewed a couple of people who have been on the front line and in the jail cell for

---

[143] Yes, there are dudes older than the President of the United States (as of this writing, July 2024). Despite news reports to the contrary, being 80-something isn't fatal. Case in point: Mick Jagger.
[144] Revolt of the Aging Ragers would make an excellent band name.

participation in protests recently. To be taken with several grains of salt,[145] I offer....

TIPS FOR AGING RAGERS

1. Investing in one of those canes that converts to a stool is not a sign of conceding to degeneration. These are effective tools for tripping fools and anyone suggesting that "granny should be at home," if wielded with stealth. You may wish to practice this maneuver after church services, targeting that heathen who takes money out of the collection plate, or, for advanced practice, outside the biker bar. Neither is recommended, of course, but we all have unique tolerances for risk. Be sure your umbrella insurance policy is paid up first.

   Executed properly, no one will come after the elder stooped over their stool. Added bonus: do not overlook the benefit of being seated, which makes you lower than the youngsters, when the rubber bullets come flying in. The kids don't have to worry about their osteoporosis.
2. Special consideration for women: while Depends diapers have their place, especially when protesting for hours on end, consider getting a "stadium buddy" (aka "female urination") device instead. Not only does it allow you to relieve

---

[145] Which may or may not rim your margarita glass.

yourself at will, but you can send a message when you do. Recall that famous scene in the movie, *The Full Monty*, when the dude was horrified to witness women peeing while standing up. Claim your power. Look that restraining officer who's trying to cuff your arthritic wrists in the eye when you let it all go.

"I tell you, when women start pissing like us, that's it. We're finished, Dave. 'Extincto.'"
The Full Monty. 1997.

---

3. Wear long sleeves, long pants, sunglasses (UV protection recommended) and a large hat. Something you might wear at the Kentucky Derby, say, or perhaps a sombrero, depending on your style of subterfuge. These help hide your identity when the drones fly over. No need to alarm the family members, especially the ones who hold your Power of Attorney, by being spotted on the news. This attire also serves as essential protection from the sun, and we've got enough carcinomas already.
4. Covering your eyes and skin will also protect your eyes and skin from the tear gas *powder*. Yes, I said powder. Did you know that "tear gas" is a fine spray of crystalline powder? Is anything what its name says anymore?? You want that stinging tear gas powder to stick to your glasses

and clothes, not your eyes and papery skin. Do consider packing a change of clothing because that powder sticks in the weave of fabric like bacterium glue.[146]

5. Pack some alcohol-free moist towelettes to clean that tear gas powder off your skin in case there's no running water available.[147] Note that the burn "only" persists for 20 or 30 minutes, but at our ages, those are precious minutes ticking away on our end-of-life clocks, so try to make the best of it by humming favorite tunes (see tip #6).

6. Memorize the words to your favorite songs. If you are hauled off to jail, you'll have no phone, no word search puzzles, and no "how to break out of jail" books with you. They can't take your voice away, so this is a good time to sing at the top of your lungs. Listen to that oldies station or cue up the '60s playlist on repeat for a solid week before the protest, so the words will be fresh. Or just wing it because what's more annoying than some old fart belting out the wrong words to an old standard? Just might get you freed sooner.

7. Know that police officers often use zip ties,[148] not handcuffs, for group arrest situations. Claim that

---

[146] The stickiest stuff naturally occurring on Earth is secreted by *Coulobacter crescentus*, a water bacterium. With the adhesive force of nearly five tons per inch, that bacterial glue could be used to lift several cars at once. WHERE IS THIS MATERIAL WHEN I CAN'T EVEN GET SUPER GLUE TO MEND MY BROKEN MUG?

[147] Extra credit for packing biodegradable bamboo wipes. Most "regular" wet wipes contain plastic fibers, potentially harmful parabens, the evil phthalates, and other chemicals. May our activism not add to the nightmares we're protesting!

[148] As if we don't already have enough plastic trash!

arthritis, rotator cuff tear, or vibrator-induced wrist pain[149] loud and proud when they try to zip-tie your hands behind your back. You can still scratch your nose itch or flip someone off, depending on your mindset, if your hands are zip-tied in front.

8. Keep those wrists far apart wherever they zip-tie you, so the hard plastic doesn't cut into your thinning skin. Gloating that you got the special front-cuff treatment is perfectly acceptable. How many times have the youngsters smirked when you tried to understand the TikTok??

9. SHUT UP. Resist that amiable elder urge to make friends with the nice officer. Remember the Miranda warning: "anything you say can and will be used against you." When asked if you have grandchildren, the old "my memory is not so good" comes in handy. Do not let them even suggest that your little grandbaby lovebugs would want you to be safe at home, instead of standing on this protest line. Of course, the "I want to see my lawyer," verbalized on repeat, is always a good mantra in just about any circumstance.[150]

10. If there's a chance you'll spend the night in jail, make your bottom-most layer of clothing comfy

---

[149] If *Grace and Frankie* (Netflix show starring Jane Fonda and Lily Tomlin) taught us anything, it's that wrist soreness from vibrator use is a real issue for mature women. Get guidance on safer products from AARP. Seriously. See their "Where to Buy the Best Sex Toys" article. https://www.aarp.org/benefits-discounts/members-only-access/info-2024/in-the-mood-buying-sex-toys.html

[150] The one obvious exception being during sex.

for sleeping. You may be restricted to wearing a single layer of your clothes as jailors often demand you remove any top articles of clothing. Wearing your "Defund the Police" tee as a base is not recommended.

11. Write down phone numbers of family members and your family lawyer or the local chapter hotline of the National Lawyer Guild (NLG) in Sharpie on your arm. If arrested, your phone will likely be taken from you, so this spares you the agony of trying to remember the gosh darn numbers because who knows anyone's gosh darn phone number anymore?[151]

12. Prep a cardboard sign with your favorite rant, both to get your message heard but also as a first line of defense against the rubber bullets. Consider the size of your signage carefully. A small sign will be easier to tote, sparing precious energy. A larger sign will block more rubber bullets and tear gas . . . and your face if preferring not to be identified.

13. Take lots of pictures on your phone or old school camera, as evidence of the peaceful nature of the protest (reassures the family!) and/or abuse-of-power incidents (protect your rights!). Resist all temptations to share those grandbaby pix, however. Remember note #9.

---

[151] I can tell you the home number of my youth (436-7640) but couldn't tell you my best friend's number today.

14. Goes without saying: bring water, medications, masks, sunscreen, phone and an external charger, those compression socks, knee brace, back brace, wrist brace, all the braces, and yes, your identification. Tempting as it may be to remain anonymous, lack of identification can delay your processing at the jailhouse by hours. Your bladder urges you to speed things along.
15. Speaking of remaining anonymous, do confide the time, date, and location details of your outing with someone who has the time and the wherewithal to bail you out of jail. Preferably someone already sworn to keep your secrets because you "know too much."
16. Most importantly, stock up on whatever indulgence you'll enjoy upon your safe return home. A fermented beverage, perhaps, or any of the foods on the "forbidden foods" list from that nutritionist who is hellbent on sucking all the joy out of your life. You've earned it.

GET CREATIVE

Stumped on slogans for your signs? Consider these:

1. Protect our Planet. Secure our children's future.
2. My arms are tired from holding this sign since the 1960s.
3. Make Earth COOL again.
4. There is NO Planet B.
5. Not fragile like a flower. Fragile like a BOMB.
6. Stop DENYING our Earth is DYING.

7. Our granddaughters should not have to fight the battles their grandmothers already WON.
8. System change, not climate change.
9. And still, she persisted.
10. Burning fossil fuels today incinerates our children's tomorrow.
11. Care for Earth like it's your HOME. Because it is.
12. A woman's place is in the Resistance.
13. Seen on an embroidery piece by a craftivist: "I'm so angry, I stitched this so I could stab something 3000 times."
14. Act now or swim later.
15. I went to Planned Parenthood and all I got was a breast exam, a pap smear, STD testing, cancer screenings, a pregnancy test, and access to affordable birth control.[152]
16. I DISSENT.

BUT SERIOUSLY, FOLKS

Though this chapter contains some legit suggestions from those in-the-know, it's also a tongue-in-cheek summary. I urge you to check out these legitimate resources for tips. Preparing for activism at any age is serious business.

---

[152] I can vouch for this. Before I had health insurance, I went to Planned Parenthood to receive all these services and never had a single abortion there.

From the **Human Rights Campaign**:
https://www.hrc.org/news/tips-for-protesting-peacefully-and-safely

From the **National Lawyers Guild**:
https://www.nlg.org/massdefenseprogram/protest-tools/

From AARP:
https://www.aarp.org/politics-society/advocacy/info-01-2009/how_to_be_an_effective.html

---

"The history of our country is that nothing happens until people start putting their bodies on the line and risk getting arrested."

Ben Cohen of Ben & Jerry's Ice Cream

SIXTEEN
Playlist Track 7: Ocean Acidification

♫ **Recommended Musical Pairing** ♫
"Only the Ocean"
Jack Johnson

Good news, folks! The "Earth Out of Bounds" study says this boundary has not yet been breached, but we are bumping up against it. Which is good news like your doctor saying you aren't obese YET, but one more cheeseburger could do you in.

Oceans absorb a good chunk of the CO_2 we put into the atmosphere, currently about 30% of our global CO2 emissions. Great as that sounds, the increasing CO2 levels make the waters more acidic—about 30% more since the Industrial Revolution began. Which is not good news for coral reefs, shellfish and phytoplankton. More acidic waters hinder the formation of calcium carbonate which is essential for shell and reef-forming.[153] If the waters become too acidic, their shells could begin to dissolve.

Oysters, mussels, scallops and coral reefs may be worse off than crabs and lobsters but they will all be affected. And they are already overstressed by all the pollution in the waters from agricultural and urban runoff.

Well, so what, we'll just give up eating shellfish, cancel the snorkeling trip, and all is well, right? Let's move on, we've got bigger problems.

Unless you care about drugs to treat cancer, arthritis, bacterial infections, viruses, and other diseases, because many of those are developed from coral reef plants and animals. And who knows how many more have yet to be discovered?

Coral reefs support some 4,000 species of fish, 800 species of hard corals, and hundreds of other species—coral reefs being home to more species by area than any other marine environment. When you consider the vast size of the oceans, that's like saying, out of all the hotels, motels, rental homes, hostels, lodges, cabins and cottages, the majority of travelers stay at Holiday Inns. . . and we're about to tear them all down. As a cancer survivor, I'm Bruce Banner-turning-Incredible Hulk furious we are ruining untold opportunities to discover new treatments and cures, simply because we can't shut down the fossil fuel spigots.

---

[153] $H_2O + CO_2$ makes H_2CO_3, carbonic acid, which dissociates to hydrogen ions (H+) and bicarbonate ions (HCO3-). The hydrogen ions bond with CO3- carbonate ions, the ones that shell-making critters use to form their shells. More hydrogen, fewer carbonate ions available for making shells.

Too bad coral reefs are already stressed by the warming ocean temperatures (also a product of too many greenhouse gases), rising sea levels, and pollution. Corals depend on microscopic algae that live in their tissues for food (and for their pinkish color). When faced with big temperature changes or too much pollution, the algae vacate like a deadbeat roommate who owes you money, causing the coral to turn a pale color. "Bleached" corals that have lost their food sources starve and become more susceptible to disease.

Coral reefs also provide priceless protection for shorelines, buffering some 97% of the destructive energy from waves, storms and floods. Devastation in coastal communities from a major hurricane could double if coral reefs are destroyed, causing untold billions of dollars in damage. Read that again. Damage from major hurricanes could DOUBLE if we lose coral reefs. And that's not a good double, like two scoops of ice cream instead of one.

Ocean acidification is also alarming if you care about the world's fisheries, as many depend on healthy coral reefs. The commercial value of coral reefs to U.S. fisheries tops $100 million. Fisheries are priceless to the millions of island and coastal people around the world rely on fish for their daily diet. Not for the occasional all-you-can-eat sushi bar, but for daily nutrition. How would you feel about strangers raiding your refrigerator—if it contained the only food you have to eat?

Ocean fisheries already face a host of challenges, not the least of which is overfishing, but none as dire as the possible dearth of phytoplankton, the foundation of the marine food chain.

We have the tiny phytoplankton[154] to thank for much of the CO_2 the oceans capture. Like plants, they suck it in to fuel photosynthesis, which produces oxygen. Well, that's nice, but how much oxygen can such tiny organisms produce? Like an infant with diarrhea, a whole lot more than you might imagine: 50-70% of all oxygen produced. Can I hear a hip, hip, hooray for Team Phytoplankton? Such a tremendous service provided by some of the tiniest among us. Why on Earth would anyone want to make their jobs harder? Makes as much sense as crippling our farmers before they plant our food.

~~"We just didn't know..."~~

"We just cared more about money."

Phytoplankton also form the foundation of the marine food chain, feeding shrimp, small fish and whales alike. If they go hungry, we go hungry.

Too bad phytoplankton are one of those shell-building critters affected by ocean acidification. If they can't build competent shells, we risk losing it all. No shells mean no phytoplankton sucking up our CO_2, providing us with oxygen, and feeding countless fish species—the collapse of the marine food chain, in other words. Never underestimate the tiniest among us.

"You may have never given phytoplankton a thought, but if they up and disappeared, it'd only be a matter of time before you did too. These powerhouses put more oxygen into the atmosphere

---

[154] How small, you ask? A soda can full of ocean water would contain between 75 and 100 million.

than any other organism, form the basis of the marine food chain, and safely sequester carbon to the abyss. Their numbers have been steadily decreasing because of human activity, and this should concern each and every one of us."

"Why You Should Really, Really Care about Phytoplankton."
Green.org

Dammit. How do we keep ending up here, bad news escalating into the harbinger of eco-apocalypse.

As if the oceans aren't facing more stressors than an air traffic controller during the world's biggest software outage,[155] along comes a news report that upends even my barest hope we can stem the awful tide of acidification. As is so often the case nowadays, the worst news came from good news. Scientists have discovered oxygen is somehow being produced in the vast depths of the ocean, some 12,000 feet deep, particularly in an area called the Clarion-Clipperton Zone which spans 1.7 million square miles between Hawaii and Mexico.

Just HOW, they haven't yet determined conclusively, since no sunlight penetrates to that depth, which means that photosynthesis, the usual method of producing our oxygen, is impossible.

But get this. "Polymetallic nodules" on the ocean

---

[155] The July 2024 CrowdStrike software update caused "blue screens of death" around the world. Idled workers rejoiced across the social media platforms—all except for the IT techs who had to fix it all.

floor, containing metals like cobalt, nickel, copper and lithium, may be the source. The nodules, with diameters about the width of a small or large paper clip, produce a tiny amount of electricity, about 0.95 volts. Scientists theorize the nodules band together, so to speak, to generate the 1.5 volts needed to split ocean water into hydrogen and oxygen ions. HOW FREAKING COOL IS THAT?! Best "rock band" ever.

A myriad of questions remain, but the discovery of a previously unknown source of oxygen—that gaseous substance that most life forms require—sounds worth investigating, doesn't it?

Too bad several companies have already received permits[156] to disturb these deep-ocean environments to collect polymetallic nodules for mining the metals they contain. So far, the mining has been "exploratory," not yet ramped up for commercial operation, but could commence in 2025.

The process of mining may prove devastating for the deep-sea critters who call this place home. The mining operations create vast clouds of suspended material which may not subside for decades—and may prove deadly for deep-sea critters. And these aren't any ordinary sort of critters, but ghostly "gummy squirrels" and egglike "ping-pong sponges." There's a sea cucumber that looks like a spring roll covered in suction cups. You really must see them to believe them. In fact, Google "Clarion-Clipperton Zone species" and see for yourself.

---

[156] From the U.N. affiliated agency called the International Seabed Authority.

I'll wait. This is the stuff aliens in sci-fi movies are made of.

Poor kids have no idea what they're in for. And guess who else isn't in favor of destroying this area? Many South Pacific Island nations, including Palau, Fiji, and Tuvalu, have called for a moratorium.[157] Not content to flood their homes with rising seas, we're about to ruin the ocean floor beneath them. Are we really set on kicking people when they're already down?

"Scientists have warned that deep seabed mining will cause biodiversity loss, both by destroying seabed life where mining would take place, with little prospect of recovery, and by generating plumes, light, toxins and noise that could impact both benthic and mesopelagic marine life far beyond actual mining sites."

2019 Position paper
Deep Sea Conservation Coalition

Why this rush to mine these tiny nodular powerhouses? So we have more metals for the batteries in our iPhones and electric vehicles, of course. Since mining for such things on land tends to be an environmentally awful process, why not make "our"

---

[157] Note that not all island nations oppose the mining. Some, like Nauru, Kiribati and Tonga, seek income from permitting such projects. No judgment from me. They need funds to deal with all the climate change impacts.

messes in someone else's backyard?

But hold on, we may not even need the metals those nodules contain because the future of batteries may be salt.[158] The world's largest battery manufacturer, Chinese CATL, is working on sodium-ion batteries that could eliminate the need for mining metals like lithium. A Chinese carmaker, Chery, is already supplying their electric vehicles with sodium-ion batteries.

No, these batteries aren't in our smart phones yet, but does it make sense to destroy yet another habitat for a resource we may not need after all? Could we please take a more thoughtful approach this time around? Pause to consider the awful ramifications of the mining operations for this oxygen-producing region before we rush in to "get while the gittin's good"?

There's so much more to this story than I can relate, and given the new-ness, it'll probably be out of date by the time you pick up this book, anyway. I do recommend the "Deep Sea Mining" episode of John Oliver's *Last Week Tonight*, first aired June 13, 2024, if you want a more thorough (and enraging) overview of the issues as they are now.[159]

Some tech and car companies, including BMW, Google Volvo and Samsung have pledged not to use metals extracted from deep-sea mining—not until the true environmental impact is known. I respect their restraint. Healthy oceans make Life as We Know It

---

[158] https://youtu.be/qW7CGTK-1vA
[159] The episode is on YouTube. A comment on that YouTube video by @Ekuahx reads: "I'm a mining engineer and wrote my master's thesis on this . . . The environmental impact from this is off the charts, unlike anything we've ever seen."

possible. Isn't that worth a pause before we charge in, yet again, like those smallpox-carrying missionaries?

"Those nodules grew over millions of years and if we take them out now, we don't understand how many species depend on them—what does this mean for the beginning of our food chain?"

Claudia Becker, BMW sustainability expert

So what song am I singing here? Ocean acidification is yet another disastrous result of what we've already discussed ad nauseum: our incorrigible habit of burning too many fossil fuels. So, it's time to double down on the renewable energy projects.

I'm red-faced as I write that. Today is a scorcher and the A/C in our home is running full blast just to keep the indoor temperature at 80°F. We've investigated solar panel options multiple times, only to be told we have too much tree cover and shading from the adjacent hillside to make a solar system cost-effective.

The Los Angeles Department of Water & Power provides electrical service for our A/C and for some 4 million other folks around here. Renewable sources currently make up some 34% of the electricity LADWP provides and a horrifying 21% still comes from coal. I crumple with eco-grief every time someone in this household cranks down the thermostat. Seeing as how the sun shines some 290 days a year in Los Angeles, that's often. Unlike the cooler coastal communities, hot temps in inland areas like where I live are common some eight months out of the year.

So, we did what we could. We signed up for LADWP's "Green Power for a Green L.A." program. For an extra 3 cents per kilowatt hour, we help fund LADWP's purchase of renewable energy.

But . . . is this program so much greenwashing? The cynic in me always wonders. The most recent annual report published by LADWP, 2022-2023, summarizes amounts raised and spent on renewables. The report concludes that some 30,542 metric tons of carbon dioxide emissions were eliminated in that timeframe, thanks to the program. I'm no finance whiz, but that sounds like a good result to me. But then, I'm happy when the bottom line of our household budget is a number in black.

"The sea, once it casts its spell, holds one in its net of wonder forever."
Jacques Yves Cousteau

CITIZEN SCIENTISTS

Whether you've got a Ph.D. in biology or failed it in high school, all you need are curiosity, patience and time to become a Citizen Scientist. Folks concerned about any or all the eco-crises are volunteering to help collect samples, make field observations, take photographs, classify images, analyze data, develop technologies, and more to help those with the credentials do the critical studies. If you've got the time and interest, there's a scientific study that needs you.

Citizen Science is an official U.S. government platform to facilitate collaboration between citizen scientists and the federally supported research projects. As of this writing, the site listed 503 projects seeking citizen scientists in the Chesapeake Bay area, Hawaii, Alaska, and the Florida Gulf Coast, as well as nationwide and online.
https://www.citizenscience.gov/

NeMO-Net Video Game. Concerned about coral reefs? Trick question; we all should be. Good news, there's a game for that. You can help NASA classify coral reefs by highlighting on 2D and 3D images scanned from the ocean floor. These are sent to NASA to help teach their supercomputer to classify coral reefs on a global scale. Available for iOS and Windows devices. Who wants to play?
http://nemonet.info/

NASA Science. Here's your chance to claim that rocket science credential, folks! NASA needs your help in examining space telescope data, inspecting images to find comet-like objects, and listening to radio signals to search for intelligent life beyond Earth, and more. Not restricted to the U.S., these projects are open to people around the world.
https://science.nasa.gov/citizen-science/

iNaturalist. Through this app, observers record encounters with tracks, nests, carcasses and critters at a particular time and location. These records help build databases of biodiversity and help scientists identify species and areas in need of protection. Oh, and it's also helpful for those "what's this I planted here and can't remember the name?" questions. Ask me how I know.
https://www.inaturalist.org/

Society for Science. This nonprofit has been dedicated to bridging the gap between scientists and citizens, expanding scientific literacy and supporting research for more than 100 years. Founded because "a healthy democracy depended on a public understanding of science," the Society offers support for young scientists and their educators in learning to explore the world around them with rigor and professionalism.
https://www.societyforscience.org/research-at-home/citizen-science /

SEVENTEEN
From The Standup Stage: So Woke

Hello, my name is Cheryl. My pronouns are she, her, and goddess.

And I am a woke-aholic.

It's been [checking watch] 19 minutes since I said something woke. BLACK LIVES MATTER!

Ooops, make that ZERO.

This is a safe space here, right?

I admit I'm powerless over The Woke.

GET THIS: I've been known to give water to people waiting to vote on a hot day. And I've taken my impressionable children to drag queen shows. I have woke fantasies about the Queer Eye Fab Five. Deep soulful conversations with Karamo. Jonathon styling my thinning hair, then saying, *"Look how gorgeous you are!"*

SO HOT. [bending over to catch breath]. Hang on, I need a minute. [standing up, fanning face]. Because that's what us WOKE WOMEN WANT: someone to listen to us, really listen to us! Not the "yes, dear" crap. And if they're my age . . . we want someone to make our wrinkles disappear.

My ideal dinner companions would be RuPaul, Kamala Harris,[160] Dr. Fauci, and Ron DeSantis. One of these does not belong? Well, we are going to need somebody to tell us if we're woke enough! If Ron's not on fire, then we have to try harder.

See, I didn't even know I had a problem. They say that's the first step, right? Admitting you have a problem? I really didn't know I was WOKE until Ron DeSantis came along and set me straight. After he said, "Florida is where woke goes to die," I realized.

I CAN NEVER VISIT FLORIDA AGAIN. I can't risk it. I've paid into my life insurance policy far too long, and it might not even pay out. Somebody like me going to Florida, they'd say I was asking for it and call it a SUICIDE.

But living here in my "sanctuary city," I'm just flaunting my wokeness. I am a longtime subscriber of NPR, and I have the tote bag to prove it. I carry pictures of Terry Gross in my wallet. SO WOKE.

I believe women are entitled to lifesaving healthcare and autonomy over their bodies. Can you imagine the outrage if a man was going through a painful experience, trying to pass a gallstone, say, and the doctor said, "We have to wait to see if you're gonna DIE before we treat it"? Because if you've ever had a gallstone situation, you know—you're already SURE you are going to die.

---

[160] Whom I pray is the president of the United States by the time you read this. If not, consider me permanently drunk.

I also believe people should be able to get gender-affirming care if they want it. I can't imagine how awful it would be to feel trapped in a body that doesn't reflect who you think you are. Welllllll.... [glancing down at pear-shaped middle] every day I get older, the more I can relate.

Oh, and I've saved the WORST for last: I don't "believe" in climate change. Nope, I don't. Because I KNOW it is happening. It's a matter of fact, not faith. I studied it in college way back in the '80s, back when nobody said the ice caps are melting because of hair dryers.

Nowadays, I write about something I call eco-mindfulness, finding deep spiritual meaning in encounters with injured pigeons, spitting squirrels, mating emu videos, and a compost pile. DOES IT GET ANY MORE WOKE THAN THAT?

And I'm not even sorry. Not even trying to cover it up. See, I published a book called *Love Earth Now*, which is obviously a dog whistle, designed to recruit other libtards. Yep, I'm the lowest of low: a woke groomer.

[Checking watch] Oh, look at the time. I've gotta drive my Prius to my woke book club. We meet over at the Ruby Fruit...that lesbian wine bar in Silverlake? Yeah, we're reading all the top banned books and assessing their level of wokeness. If a book gets a high score, we buy extra copies to send to certain governors' and Senators'...CHILDREN.

Bwahahahahahahahahahahahaha!

BECAUSE THAT'S THE KIND OF HORRIBLE SHIT US WOKE PEOPLE DO.

> *"If I did not laugh," he said, in response to the criticism he received for seeming to make light of apocalyptical events, "I should die."* [161]
> *-President Abraham Lincoln*

---

[161] Want to learn more about how Lincoln relied on humor to get him through the apocalyptic challenges of his time? See "Stay Cool" in Resources.

EIGHTEEN
Playlist Track 8: Air Pollution

♪ **Recommended Musical Pairing** ♪
"Pollution"
Tom Lehrer

The next not-yet-breached boundary in the "Earth Out of Bounds" study is "atmospheric aerosol loading," which, in case you were wondering, too, has nothing to do with hairspray. The "aerosols" in this case are air pollutants; specifically, the particulates in the air resulting from both natural occurrences, like wildfires, volcanoes, and dust storms, but also from human activities, such as burning fossil fuels, coal, dung, peat, and firewood.

If it occurs naturally, why is it a problem when people do it? Is this a "do as I say and not as I do," Planet Earth?

Let's first consider some basics. The good news about particulates is that, unlike those guests who never leave, they don't linger for decades like greenhouse gases. A good rain will wash some out of the sky, which is why we see blue skies here in Los Angeles after a downpour. Nature had a plan for cleaning up her messes.[162]

The problem arises, as is so often the case, when humans take a good thing and throw it out the window. We have added so much more particulate pollution that we can't count on rain alone to clean it all up. For one thing, rain doesn't fall uniformly around the world or around the calendar year. And rainfall is getting less predictable all the time.

An "atmospheric brown cloud" now forms over the Indian subcontinent every dry season between October and February. The result of burning fossil fuels, dung and wood, plus industrial emissions, it causes severe health problems, such as asthma, bronchitis, irregular heartbeat, and even heart attacks and strokes. A 2002 study estimated some two million people die each year from conditions related to the brown cloud.[163] Same as joining a gym doesn't equal fitness, waiting for enough rain to fall is not a solid plan.

Like bad news gone worse, the brown cloud also reduces the amount of sunlight that reaches Earth's surface, decreasing evaporation and rainfall, making droughts worse. . . and less frequent rains to clear the skies. In fact, the "Earth Out of Bounds" study concluded that the aerosol loading boundary has been breached in the high-pollution areas of southern Asia.[164]

---

[162] I can think of some plastic-producing people who should follow her lead.
[163] Yes, that was in 2002! Imagine what that number must be now.
[164] Lest we get too sanctimonious about our relatively cleaner skies in the US, I wonder how much of that Asian air pollution results from manufacturing the cheap crap we Americans insist on ordering instead of supporting local suppliers (if we can even find any).

Not all aerosols are created equally, however. Light-colored particulates reflect sunlight which can cause cooling. In fact, this just in from the No Good Deed Goes Unpunished Department: cleaning up the emissions of super tankers may be causing an awful result. Shipping fuel regulations passed in 2020 have reduced sulfuric dioxide pollution by some 80%. Sulfur dioxide is a major pollutant causing acid rain and major human health problems, from respiratory illnesses to lung disease. So, reducing sulfur dioxide emissions from all those tankers that deliver the crap we order from Amazon and Temu, that's good news, right?

Yes and no. All those sulfur dioxide emissions formed clouds of light-colored aerosols which reflected a lot of sunlight. Reducing those emissions means fewer sunlight-reflecting clouds, which, dagnabbit, means more sunlight hits the oceans, causing them to warm even more. If you've read the Ocean Acidification track, then you know. . . that's not good news.

Darker colored particles, "black carbon" (think "soot"), on the other hand, form when burning fossil fuels, biofuels like peat and dung. Diesel engines, cooking stoves, wood-burning furnaces and forest fires top the charts as the biggest producers, however. These all send particulates into the sky which produce a dark residue when they fall back to Earth. I'm picturing a bleak scene from London in a Dickens novel where clouds of soot enshroud the orphan waifs.

Those falling particulates pose a big problem when they land on light-colored ice and glaciers which would otherwise reflect most of the sunlight that reaches them.

As any Goth attired in all black on a scorching day can attest, dark colors absorb light, converting it into heat. Those heated-up black carbon residues are likely hastening the melting of Arctic ice. Cue the starving polar bear sobs.

Inhaling black carbon is bad for human health, too. It's a known carcinogen that can cause lung cancer, heart disease and stroke. Exposure to black carbon causes millions of premature deaths each year. Black carbon is also considered the second largest contributor to climate change after carbon dioxide.

Controls on pollution sources in places like North America and Europe have significantly reduced the production of black carbon from smokestacks and tailpipes. Burning biofuels without any emission controls occurs more often in Asia, Africa and Latin America—places where, I am assuming, people lack affordable access to emission control technology. The double whammy writ large.

So what song am I singing here?

I'm fretting about my fireplace habit. I'm red-faced to admit that I dearly love a roaring fire in living room fireplace on a cold night.[165] Fire sparks my imagination. I can sit and stare at the flames, just noodling and imagining for hours. It's all great except it's not great for the air here in smoggy Los Angeles. There's no emission-reducing insert in our fireplace, so I'm contributing to "atmospheric aerosol loading," too.

---

[165] "Cold" being a relative term, but anything below 50°F in my case.

I tell myself it's okay because I'm such an infrequent contributor. I never light a fire on a No-Burn day—when the local air quality district issues a ban on burning wood in residential fireplaces and stoves because of already-high levels of particulate pollution in the air.[166] But can I still enjoy that fire now that I know about black carbon?

I've put an emission-reducing fireplace insert on my long Wish List of eco-friendly gadgets I covet. It's at the bottom of the list, I confess, because it's a luxury item, when you consider how rarely I'd use it—or need it living in a place where winter lows rarely drop below 40°F.

For now, I'm instituting stricter protocols for deciding when to light that match. First, I'll ask myself, really grill myself, how badly do I need the kind of therapy that only soot-producing wood burning provides?

And I'll pay penance for each fire I decide to light by forcing myself to clean out the cat box each time I do. Because that's a surefire way to clean the air, in our household, anyway.

CLEANER STOVES

Organizations like the **Clean Cooking Alliance** and the **Modern Cooking Facility for Africa** help people in highly polluted areas like India and sub-Saharan African obtain and learn to use biogas-fueled cooking stoves. They can be fueled by the off-gases from composting food and agricultural waste, which are collected by a

---

[166] There are exceptions for homes above 3,000 feet in elevation and for homes that rely on wood burning as their sole source of heat.

biodigester, making it an essentially free energy source (after equipment costs are paid).

Such sources remain readily available when the electric service is inconsistent—or nonexistent. They also free up many hours spent searching for and harvesting wood for fires—tasks usually undertaken by women and children.

While the biogas stoves are not emission-free, they do produce considerably less black carbon and fewer greenhouse gases than cooking over wood, charcoal, animal dung or even kerosene. Burning such fuels also produces dangerous toxic smoke with devastating health effects for the one tending the stove.

https://cleancooking.org/

https://www.moderncooking.africa/

Solar Cookers International assists people around the world to obtain and use solar-powered stoves. This makes for another free source of energy, one that's free of greenhouse gases.

"Solar cooking empowers women and girls worldwide by offering a clean, sustainable alternative to cooking with polluting fuels like wood and charcoal. In regions struggling with energy poverty and environmental degradation, solar cooking not only reduces emissions— it saves lives, improves health, and protects our forests. By supporting Solar Cookers International, you're

helping communities and the planet thrive. Visit https://www.solarcookers.org to learn more."

Caitlyn Hughes, Executive Director,
Solar Cookers International

"Even in dark times, we not only dream, we do. We not only see what has been, we see what can be. We shoot for the moon, and then we plant our flag on it. We are bold, fearless and ambitious. We are undaunted in our belief that we shall overcome; that we will rise up.."

United States Vice President[167] Kamala Harris

---

[167] She's VP as of this writing, and I was dearly hoping you could strike the word "Vice" when you read this, but alas . . .

NINETEEN
Eco-Grief Therapy: Step-By-Step Guide

CONGRATULATIONS on your purchase of this first-of-its-kind Jigsaw Puzzle for Eco-Therapy. It's your key to calm when the world around you is a puzzling madness.

Savor the ecstasy of being able to SOLVE something for freaking once!

WHAT YOU'LL NEED:

- One Jigsaw Puzzle for Eco-Therapy
- Baking sheet (optional)
- Whatever shred of patience you can muster

NOTE: YES, this is standard, rectangularly shaped puzzle, unlike the climate crisis which is as amorphous as a box jellyfish[168] and its sting, just as lethal.

---

[168] "Jelly-fish." What an innocuous-sounding misnomer for such a deadly creature. The Australian box jellyfish is considered the most venomous marine animal. The venom from any of its fifteen ten-foot-long tentacles "turns the tissue into soup," causes the heart to seize, and can bring death within just four minutes.

TERMINOLOGY NOTES

- **corners:** pieces that have two solid edges meeting at a right angle. For your "piece of mind," this puzzle contains four corner pieces.
- **edges:** pieces that have one solid side—no innies or outies.
- **outies**: the protruding part on any side of a puzzle piece, may also be called knob, key, loop, or male.
- **innies**: the dented part on any side of a puzzle piece, resembling the hole in your heart, may also be called the blank, hole, lock, socket, or female.
- We elected to use the terms "innies" and "outies" in these instructions because thinking of belly buttons gives us the giggles.
- **shithead pieces**: they lack any straight edge. Their four sides of randomness give you no respect, not even a clue where they belong in the greater scheme. Allow the rudeness of these shithead pieces to remind you of the lack of respect Earth has suffered, thanks to the heartlessness of humanity. Use that fury to fuel you forward.

STEPS

1. Dig out your largest baking sheet from the kitchen cabinet—unless it sits in the sink covered in grease, then skip this step. This is not the time to clean up yet one more freaking mess.

2. Clear the kitchen table, or any surface that's easily freed of the daily detritus. Use a corner of the kitchen floor if no table is available, brushing aside the crumbs from any cracker-smashing spree[169] before sprawling.
3. Open the jigsaw puzzle box and dump the pieces on your baking pan—this will help corral the runaway pieces, giving some bounds to the task at hand. Or scatter them willy-nilly if you're fed up with the freaking "boundaries" that hem you in.
4. Yep, it's a mess, just like the environment that sustains us. Notice if the jumbled mayhem triggers any frustration, anger or despair. Does the chaos, the lack of clarity in the "picture" make you feel like quitting? If so, dare to stand in solidarity with every jigsaw puzzle builder—and Earth lover—across time in trusting that some sort of coherence will evolve. Or curse the whole mess and walk away for today. You do you.
5. When you're ready to continue, turn over all the pieces so that the printed sides face up. Yes, this takes time and patience which may tap your near-empty reserves, given the fossil fuel companies' reports of record-breaking profits. Allow some days (weeks, months) for this step, if this simple act of "righting" feels too hard, too

---

[169] Smashing crackers provides satisfactory therapy in a pinch, as long as you don't have anyone in the house who's gonna complain about the mess. You do not need that nitpicking noise right now.

ECO-GRIEF THERAPY: STEP-BY-STEP GUIDE

forced. Don't rush yourself, and don't censor yourself either. Curse at will.

6. Now separate out the edge pieces, featuring a straight edge along one of the four sides. Express any gratitude you feel, to receive even these smallest hints of guidance. Set aside the more sadistic pieces, the shitheads without a straight edge, the ones that refuse to give you even a hint of where they belong—much like the shattered pieces of what used to be considered "normal life."

7. Here's where the puzzle building process turns ugly. Do your best to sort the edges into groups of similar colors and patterns on your building surface. Express any frustration with multicolored pieces for which no one category suffices by pounding the table or yelling at the cat. That's what they are there for.

8. Experiment with connecting two pieces, inserting outies into innies, while keeping your mind out of the gutter, if you're building the puzzle with children present. Notice any feelings of frustration or fury that arise if not a match. Pounding ill-fitting pieces in punishment is acceptable and may even be necessary for continued work.

9. Notice how this act of building, making a solid line, bringing some sort of order to chaos, even if punctuated by pounding, feels. Consider the emotions this process evokes—unless you're sipping the cooking sherry you found when you

went looking for a pan and now feel nothing at all.
10. Continue assembling the edges of the puzzle, taking breaks to smoke, to meditate, to cry, or to laugh maniacally, as needed. But avoid Facebook, Twitter, Instagram, Google Photos, your messaging app and anywhere else you risk encountering depictions of people going about their oblivious lives while the world burns.
11. Accept that the solid edge you are building will, inevitably, include one or two disruptions because some pieces are always missing. Despite our superior quality control, there are always missing pieces, much like the missing answers to resolving climate denialism. Don't give those missing pieces the satisfaction of ruining your sense of accomplishment. In fact, give yourself a pat on the back for catching on to their wretched trickery. You know what's up, even if you don't know how to fix it.
12. It may be time for another break. If you gave up smoking years ago, this is not the time to consider restarting, tempting as it may be.
13. When you've completed the border—or you've given up on it entirely, following the lead of certain legislators—consider the remaining shithead pieces. Sort them into groups based on color, pattern or whatever comes to you, taking into consideration your frustration tolerance and how much of that sherry you've had.

14. Begin connecting the shithead pieces to the edge pieces, and then to each other, listening to an encouraging podcast[170] or "Don't Think Twice, It's Alright" by Bob Dylan, as needed, to keep going. Or maybe "Everybody Hurts" by R.E.M. if you need to remember that you're not alone. But avoid Adele at all costs.
15. When your puzzle is complete—or whenever you decide you're done with it—give yourself a big pat on the back for making it this far. Savor the rare feeling of having solved some part of a real effed up quandary, to have made some pattern out of chaos.
16. Then fling the entire thing onto the floor and walk away. Somebody else will clean up this mess, right? That's what the plastic manufacturers do, destroying every aspect of the biosphere without a single apology, and you can, too.

"There are no extra pieces in the Universe. Everyone is here because he or she has a place to fill, and every piece must fit itself into the big jigsaw puzzle."
~Deepak Chopra

---

[170] Anything with Oprah or Deepak Chopra should work unless you just can't take another pep talk. Angry silence is always an option.

TWENTY
Playlis T Track 9: Ozone Layer

♫ **Recommended Musical Pairing** ♫
"That's Cooperation"
Sesame Street

Sit down, children, and let me tell you about another time when a scary environmental thing happened. Waaay back in the 1970s, scientists discovered that the use of chlorofluorocarbons (CFCs) in refrigerants used for cooling and propellants for hairspray was causing a hole in the ozone layer[171] over Antarctica. The ozone layer is that miraculous region in Earth's atmosphere that keeps dangerous solar radiation from reaching the surface and killing us. That "hole" would keep getting bigger, exposing everyone on Earth to more and more solar radiation, unless we stopped using CFCs. So, of course, a

---

[171] Technically, not a hole but a region of "exceptionally-depleted ozone" in the stratosphere.

disinformation campaign was launched, skeptics ridiculed the scientists, and nobody did anything about the scary thing, right?

NOPE! Hold your horses, pardner. Back in those Olden Times, aka the late 1980s, countries around the world negotiated and promulgated a *solution*: the Montreal Protocol. Ratified by 198 parties, all of whom agreed to phase out the production of CFCs and other problematic compounds.

"Perhaps the single most successful international agreement to date has been the Montreal Protocol."

Kofi Annan
Former Secretary-General of the United Nations

New refrigerants and propellants were formulated, and hairspray was saved. Not to mention air-conditioning, which is becoming ever more essential for surviving a warming planet.

Everybody celebrated this rare, but essential cooperative win for all of Earth, and then promptly forgot about it. As with most things the size of the entire atmosphere, however, the problem didn't vanish overnight, even if our memories of it did.

Let's refresh our seventh-grade earth science lessons for a moment. Ozone (O_3), which consists of three oxygen atoms, collects in a layer in the stratosphere some

9.3 to 18.6 miles above Earth's surface. The ozone layer absorbs and blocks most of the Sun's damaging ultraviolet radiation from reaching the surface. It protects us from dying of skin cancer—if we could live at all. Most plants would die without the ozone layer because photosynthesis would become near impossible, and that's the basis of our entire food chain, folks. No plants or phytoplankton, no fish, no mammals, no Big Macs or Happy Meals.

Gauging the success of the Montreal Protocol has been a challenge. Despite our collective "the ozone hole problem was solved years ago" delusions, CFCs persist in the stratosphere for a long time. In other words, the ozone depletion problem didn't stop just because some people signed a piece of paper. Color me shocked.

The "ozone hole" is still a thing, but it's actually a seasonal occurrence, not to be confused with hay fever season. What's more, it's not an actual hole, but a region above Antarctica where the concentration of ozone thins by a lot.[172] This thinning occurs when the weather warms in the southern hemisphere, August to October. It closes around January each year. It's variable year to year, however, trending smaller in warm years, but growing larger in cooler years.

Recent studies show that the amount of ozone-depleting gases in the stratosphere are diminishing but won't return to pre-1980 levels for a couple more decades. DECADES.

---

[172] Thinning also occurs above the Arctic Circle, but not nearly as much as over Antarctica.

> *"When it comes to a clear sign the ozone hole is going away, it could still be a couple decades before we can look up and say it's smaller every single year than it was in the early 2000s . . . It will be a long, bumpy road, but we're headed in the right direction. We just need to be patient and keep up the good work."*
> Susan Strahan, NASA Scientist

If only we could muster the resolve of all those countries back in the 1980s to solve any one of the scads of environmental crises we still face today. Not just the knowledge, the framework, or the awareness, but the sense of urgency worthy of the truly frightening calamities that we are handing off to our kids and grandkids. What better legacy could we leave? Those words brought to you by Captain Sarcastic.

So what song am I singing here?

I signed up to receive notifications from the "**We Don't Have Time**" (WDHT) movement, said to be the largest social media platform dedicated to climate action. With over 100,000 registered participants and a social media reach of over 200 million, it's the ultimate peer pressure system[173] for demanding companies, governments and other organizations to act for climate. What a relief to find such positive use for social media, considering all the negativity that it foments. Finally, a platform that gives us a chance to act with a sense of

---

[173] My words, not theirs.

URGENCY. Because, truly, we don't have time to f*ck around.

> **We Don't Have Time**
>
> The platform serves three basic functions:
>
> **To inform.** WDHT consolidates essential, fact-based information from trusted sources, so that all of us—not just the experts or pundits—have access.
>
> **To invite reviews.** Members of the WDHT community can send "Climate love" to encourage great climate action, "Climate ideas" to suggest better ways of doing things, and "Climate warnings" to urge the end of not-so-good practices.
>
> **To nudge.** Once enough reviews have been compiled, WDHT forwards them to the concerned parties to engage in dialogue and to encourage positive change.
>
> Compiling the many reviews also reduces the impact of the few wonkers whose opinions sometimes receive an outsized response.[174] Because the information is the fact-based and current, we're not wasting time arguing over decades'-old science, rumors or outright fabrications.

---

[174] See, for example, book banning....

Note that WDHT also initiated the #MoveTheMoney campaign mentioned in the Climate Change chapter.

And, yes, there's an app for that. Available in the App Store and in Google Play.

"What we're trying to do is to democratize knowledge about climate solutions and inspire and mobilize global action toward a prosperous, fossil-free future. This is why everything we do, and all the content we produce from the international climate meetings and summits, is available for free.

One of We Don't Have Time's strengths is our diverse network of scientists, activists, policymakers, CEOs of international corporations, NGOs, journalists, artists, and concerned citizens. Thanks to our reach, we can help amplify all of these diverse voices and speed up change."

Markus Lutteman, Editor, We Don't Have Time

https://app.wedonthavetime.org/

TWENTY-ONE
I Get It: I Ridicule Me, Too

I just read, with rabid attention, the *Year of No Garbage* book, and it was a bit like looking in the mirror. The book chronicles author Eve Schaub's ardent and time-consuming efforts to avoid producing trash—especially the plastic kind—for an entire year. With wit and candor, Schaub details how she scoped out eco-friendly alternatives to the nonrecyclable, nonreusable and nonsensical products we in the so-called developed world consider essentials, from tooth tablets to period panties. She has done the research, and she has the receipts. If only I possessed the concentration and focus to delve into a single topic for longer than, say. . . where was I?

I found myself on every page, stumbling along with Eve's best efforts to rid her home and daily living of senseless waste. I recognized the rabbit holes, the clinging to hope, the desperate searching for some product that would work as well as the chemical-and-plastic-laden stuff—but won't kill off Life as We Know It. The endless search for sustainably produced goods wrapped in packaging that won't persist eternally as nanoplastics for my grandchildren to clean up. Frantic

Googling for the miracle goods that will transform me from eco-vampire to environmental savior. *"She single-handedly saved us by switching to a bamboo toothbrush!"* A lunatic Earth lover can dream.

I know the looks. The smirks.[175] The ridicule, derision and scorn directed at people like me who aren't content to let Proctor & Gamble, ExxonMobil, and Amazon tell us what we should buy.

Not anymore, anyway. I do remember the days when my grandparents valued their relationship with a brand. "*If it's Smuckers, it's got to be good.*" Back when it meant something. Or we were naïve enough to believe it did.

"Chesterfield cigarettes are just as pure as the water you drink."[176]

Nowadays, I look at advertised brands with the cynicism of my college biology prof who said, "If it's advertised on TV, it's bad. Why else would they have to pay to sell it?" Advertising has changed a lot since my college days, but I think it's fair to say it's not become more truthful or transparent. Heck, I can't even figure out what they are advertising half the time. Cue the erectile dysfunction medication commercials that seem to advertise moonlit walks on the beach with a platonic friend.

Sure, the U.S. Food and Drug Administration now requires drug manufacturers to disclose the risks of their

---

[175] I've raised two teenagers, after all.
[176] Published in 1933 . . . where? *The New York State Journal of Medicine.*

medications in advertisements. So many, in fact, that I suspect we've all become numb to the laundry lists of possible effects that once seemed hair-raising.

"Side effects of this drug used to treat depression include nausea, dizziness, hives, listlessness, blue lips, loss of fingertips, prolonged hiccups, impulsive gambling, hairy tongue, fishy body odor, growth of hair on forehead, uncontrollable sobbing, and depression."[177]

What I have not seen disclosed are the devastating environmental and social impacts of the food, cleaning products, cars, or clothing advertised on TV, in apps, on social media, on billboards, on gas station screens, billboards or bus benches. Frankly, I'm surprised I haven't seen any ads broadcast on my bathroom mirror. Yet.

Where's the disclosure in fast-food commercials that cheap hamburgers are destroying the Amazon? If you read the Deforestation chapter, you know. If you didn't, I don't blame you.

Or that the palm oil in that "quick energy" snack food caused families of orangutangs to lose their homes? Countless acres[178] of rainforest have been—and continue to be—destroyed to produce the world's most widely-used vegetable oil: palm oil.[179]

Or that the Toyota Land Cruiser emits more

---

[177] These are all known side effects of different drugs; I didn't invent any of these. This time.

[178] You'll have to do your own research if you want to know exactly how many. I just can't . . . not without a serious intake of sedatives, but that makes for fairly unreliable reporting.

[179] The Roundtable on Sustainable Palm Oil, established in 2004, created a certification scheme to ensure environmental concerns are addressed. It's complicated. See "palm oil" in Resources for more info.

greenhouse gases per mile than the average commercial airplane (on a per-passenger basis)?[180] You can fight me on the math, not my strong suit, but where's the climate-killing warning on any gas-powered car commercial?

Or that trawling for the fish in your filet o'sandwich has driven the species to near extinction? The Food and Agriculture Organization of the United Nations says the number of overfished species has tripled in a few decades and that at least a third of the world's fisheries are being pushed beyond their limits.

Or that the fast fashion industry is responsible for a full ten percent of global greenhouse gas emissions, toxic pollution from poisonous chemicals and dyes, and staggering amounts of waste? A truckload of abandoned textiles is dumped in landfills or incinerated every second. EVERY SECOND. All the while the cheap prices and must-have marketing drive people to buy 60% more clothes than they did just fifteen years ago—and keep them half as long. Has any of that come up as a disclosure on any Shein or H&M ad you've seen?

Or that nearly every food product now contains plastics?[181] That warning might be difficult to formulate, considering how little the effect of all that plastic content on human health has been studied. "WARNING: This food contains plastics, but we don't know if that's bad for you because we don't care about your health enough to

---

[180] Toyota.com estimates the average emissions of the 2024 Land Cruiser to be 380 grams/mile. In 2019, the International Council on Clean Travel determined the average commercial airplane emitted 90 grams CO2 per "revenue passenger kilometer," which converts to 145 grams per mile per passenger in 2019.
[181] Is that what they mean by "special sauce"?

fund the studies." Caveat emptor.[182]

I sure haven't seen any ads disclosing any warnings, fine print or otherwise, about the environment-destroying impacts producing, consuming or using our favorite goods cause every minute of the day. Of course not, because who wants to deal with the ugly truth when we are eating our feelings? We've got enough on our...*plates*...already.

Unfortunately, I've done too much research to pretend none of that is happening. So, I beg you to give me and my desperate-to-save-the-world kind a freaking break when you snicker at our eco-friendly fixes that don't work, don't clean, don't actually save the world.

Like the author of the *Year of No Garbage*, I've done my time searching out plastic-free toothpaste, phosphate-free detergents, and compostable dental floss options. The eco-friendly dishwasher detergent left my dishes caked with food. I got my first cavity in a decade after using the chewable toothpaste tablets for just six months. I hate picking lint out of the deodorant stick that keeps falling out of the refillable container onto the floor. My efforts to corral the beeswax wrap rival my emu mating dance[183] in absurdity. My beloved bamboo forks stabs as effectively as my razor shaves when I forget to install a blade.

I even keep a small jar by my bathroom sink to collect my fingernail clippings and used floss, awaiting transport

---

[182] Latin for "the buyer is f*cked." Practically speaking, anyway. Technically, it means "buyer beware," which I guess sounds less nefarious in Latin.
[183] Sneaky reference to my book, *Love Earth Now*, Chapter 3.

to my compost bin. I'm that desperate to be able to say, at the end of the day, I *tried*. Which is wildly unsatisfying, but I sleep better.

What's more, I realize that all the emissions from shipping each individual item to me—one delivery for dental floss, another for toilet paper, another for silicone bags—mount up like the dirty dishes on Thanksgiving Day. Especially when you consider there's a Target that carries all the plastic-laden options, already delivered to my neighborhood, within walking distance of my house. Can I really justify the delivery of each individual item they carry to my house, just to save a smidgeon of plastic guilt? It's an ongoing debate among me, myself, and I. I urge you to stay clear of it.

I wish there were better options. I wish I could reach for the DelMonte can of cling peaches with the confidence my grandparents had. Ignorance just isn't bliss in these days of the Plasticene.[184]

Tempting as it is to toss in the paper towel, I just can't in good conscience concede the dream. I've already been diagnosed with uterine cancer, despite zero family history, and the genetic testing revealed no mutations. Could exposure to toxic contaminants be to blame? I have no way of knowing, any more than I know how my phone works.

But I'm determined to stick around for the kids as long as I can take the madness, I mean, survive with dignity. So, if you see me fighting with a crummy walnut hull scrubber, instead of the effective-but-disposable-

---

[184] See chapter "Playlist Track 2: Plastics."

plastic type, please give me and my compromised immune system a break. My patience is already taxed by the constant effort to refill all my refillable containers without creating a colossal mess. If I manage to eke out a few extra years with my family, it's worth it.

WHAT WE WEAR MATTERS

The fashion industry is responsible for a full 10% of global greenhouse gas emissions. I realize I've already mentioned this, but it's just so incredible we may all need to hear a few times before it sinks in.

No wonder so many of us eco-nerds now shop at thrift stores, charity shops, and online used clothing sellers like ThredUp. Given all the pollution, noxious emissions, waste, and deplorable working conditions of the fashion industry, can we please give the clothes we buy a lifetime longer than six months? For the rethreads, all the environmental and social nightmares of making our plastic-laden attire have already occurred. The sweatshop labors are done, and the trans-oceanic shipping complete. Let's honor those devotions of precious time, resources and labor as long as we can.

Because once they leave our closets, those clothes become an environmental scourge. There's so little textile recycling in the U.S., most of our used clothing either ends up in a methane-leaking landfill or on the shores of a distant land, where they may or may not be worn or recycled; many are incinerated, which creates another set of eco-disasters.

All those hordes of incoming castoffs also put local garment manufacturers out of business. Can you imagine the outcry by Levi Strauss if container ships full of jeans regularly washed up on our shores?

And how many greenhouse gas emissions are we racking up by sending our unwanted textiles around the world, anyway?

Still, it can be tough to know what brands to support, which are environment destroyers, and which employ more sustainable practices. Must I research every tag at the store before I buy a simple T-shirt? And will I even be able to afford the sustainably produced garments? I'm already exhausted from trying to figure out what's safe to eat and where to get it.

So, I turned to the podcasters, and I found a treasure trove. I'm sharing a couple for your consideration, though I confess I've only just begun to dive in. If you're interested in making more sustainable fashion choices, check these out. Or find your own. There's a lot more being said about fashion waste and sustainability than I realized, before I looked for it. Funny how that is, huh?

The Wardrobe Crisis. Hosted by Clare Press, a filmmaker and author based in Sydney, Australia, she was the first *VOGUE* sustainability editor. Her fashion bona fides are extensive, and the guests include supermodels, economists, activists, and designers. The podcast is in its tenth season, so there's a lot of info here. As of this writing, the most recent episode features a conversation with Eunice Olumide, an Afro-Scottish fashion model, environmentalist, DJ and filmmaker, about Oxfam's "Secondhand September" campaign. Why shop secondhand? Listen in.

"Never judgmental or patronizing, Press's approach to environmental activism is one we can all learn from." US VOGUE.

https://thewardrobecrisis.com/podcast

ConsciousChatter. Hosted by Kestrel Jenkins, an international journalist, innovative storyteller & conscious brand strategist; and Natalie Shehata, a multi-hyphenated and heart-centered creative working at the intersections of style and social justice, this "What We Wear Matters" podcast takes a global look at what it means to produce and purchase sustainable fashion. Conscious Chatter broadens the scope of the conversation, featuring diverse topics ranging from nonbinary upcycling artists to preserving culture as a pathway to sustainable fashion. As of this writing, the most current episode is "Why Fashion Needs to Listen to Nature When Reimagining Futures," featuring Nelson ZêPequéno, a Ghanaian-American artist, and Cayetano Talavera, a zero-waste fashion designer.

I gotta listen to that one NOW. See you in the next chapter.

https://consciouschatter.com/

"When you buy into fast fashion, no matter how many times you wash your clothes they will never be truly clean. They're stained by the sweat of those who made them, and the footprint it leaves on this planet."

Laura François, Director of Awe Exchange and socio-environmental impact strategist

---

TWENTY-TWO
Playlist Bonus Track

♫ **Recommended Musical Pairing** ♫
"Love Song to the Earth."
Paul McCartney et al

Whew. Thank you for sticking with me through this series of long, bloated chapters reviewing the "Earth Out of Bounds" report. You have no idea the gallons of caffeine required to get me through all of this—or how many times I had to stop typing to cry. I wouldn't have made it through but for thinking of you, sitting there wondering . . . will she ever get to the end of this? I forged on for you.

I assure you there was a point to this painful exercise, and here it is: to reiterate, yes, once-a-freakin'-gain, that it's not just our heretofore stable climate that's in peril. We face challenges on a host of vital fronts, and yet I hear little about any of them except the one that's spawned so much vitriol: climate change.

If you can't bring yourself to accept that human-caused climate change threatens the continued existence of people on this planet—or if it's just too complicated a conundrum for you to face head on[185]—how about doing something about the many other eco-crises?

The other purpose of this whole exercise has been to shine a light on the brave souls who haven't plugged their ears and chanted "na-na-na" and ignored another awful environmental truth, as I'm often tempted to do. For those of us sitting in our comfortable homes, it's all too easy to tune out, binge watch another series, and forget about the people drowning in floods, plastic waste and rising seas.

Why do we do that? Because figuring out what to do about any of it, what will make a significant difference, is as challenging as solving a Rubik's cube while drunk and blindfolded. After decades of struggling to figure out how to live more eco-mindfully myself, I have run into so many roadblocks and minefields that I suffer PTSD whenever I read another "Top Ten for the Planet!" list. If one more person tells me to change my lightbulbs or remember my reusable bag when grocery shopping, I will need another trip to the rage room.

> *"Climate Change is just us continuing to fuck around even tho we already found out."*
> *@casual.nihilism on Instagram 6/14/23*

What if the best thing we can do is to just NOT? We in the U.S. love a good fix, be it a glitzy new technology, or

---

[185] And I can totally understand why.

rushing in to take over and take charge. So often, however, like when your spouse is loading the dishwasher all wrong, the best thing to do is not take over. Ask me how I know.

Buying an eco-friendly doodad involves a whole host of farming, mining, manufacturing, and/or transportation, with impacts all along those pathways. Buying nothing carries none of them and doesn't further line the pockets of the oligarchs of Amazon, Walmart or Shein.

To make myself feel better about what I can't fix, I made a list of all the things I'm NOT DOING for Earth. I give myself five brownie points each, which, referring back to the "Cooking Up Therapy" chapter, means I get the whole pan to myself.

WHAT I'M NOT DOING FOR EARTH

In the name of caring for Earth, I am NOT:

- Sending my fruit and veggie discards to the landfill where they will never rot. They get to decompose in my home compost bin with the green iridescent beetles.
- Letting the water run as I brush my teeth, scrub dishes or shriek in the shower.
- Buying any fast fashion from Shein, Zara, H&M or any other brand on the "Worst Fast Fashion Brands" list.
- Sending the torn comforter off to the landfill. I mended it with my shoddy sewing skills, but well enough to donate it to a homeless shelter.

- Putting food in my refrigerator and forgetting about it until the stench prompts me to compost it. This list lover keeps an inventory of the perishable food we need to eat while it's still good, posted in plain sight. Which also reduces those annoying, "what is there to eat?" queries. There are still oversights, but when I remember that something like 40% of the entire U.S. food supply gets tossed instead of eaten, wasting so much water, labor, fertilizers, and fossil fuels . . . I redouble my efforts.
- Buying a new phone charger when I can't find mine for days. It's a small thing, but by pausing before rushing to order a new one from Amazon, I found the charger and saved myself $7.99, not to mention a slew of transportation emissions.
- Accepting that swag bag at the conference, full of stuff I do not need, saving myself the anguish of figuring out how/if I can recycle, reuse, redistribute it all . . . or feel guilty about throwing it into the trash can.
- Buying single-use batteries that I have to take to the hazardous waste roundup when they die. I keep a good supply of rechargeable batteries, in the usual AA, C, D sizes, anyway, so they can be reused for many years.
- Grabbing fast food when I am out running errands, and I'm convinced I am "starving." I wait until I'm home (turns out, I wasn't actually starving at all) and eat food I've already purchased.

- Tossing the non-working blender I got at a yard sale (when will I learn?). I took it to the Repair Café and got it fixed up to use.

That's just a sampling, abbreviated for reasons of reader disinterest. Much as I do love my lists, I am keeping your "is she done yet?" attention span in mind.

It's also a work in progress, as I've discovered it's harder than I imagined to catalog things I'm NOT doing. Henceforth, I vow to seek out the "not-do" opportunities, even if it means I must pass up that Pumpkin Spice Latte. . . .. oh, the sacrifices!

What's on YOUR list?

RESOURCES FOR NOT SPENDING

BUY NOTHING.

Buy Nothing groups create hyper-local communities where people give freely without expectation of compensation, not even a trade. Members give and ask for items or services ("gifts of time") within a community of people in a set geographic area. Not only do Buy Nothing groups keep scads of usable items out of the landfill, but they also help strengthen local communities. If the COVID-19 pandemic taught us anything, I hope it's that our local communities are there to support us when the global networks break down.

I'm the admin for my local Buy Nothing group, and it's heartwarming to see people sharing so generously, especially in these days of inflation. And saving a ton of usable stuff from the landfill warms my Earth-loving heart.

There are Buy Nothing groups in all 50 U.S. states and 128 countries. To find a Buy Nothing group near you, download their app from Google Play or the App store. Or check the website for groups hosted on Facebook and other platforms.
https://buynothingproject.org/find-a-group

FREECYCLE.

Freecycle is another nonprofit, free resource for giving and receiving items. Members join a specified Town, which may be larger than a Buy Nothing group area. They can also create a "Friends Circle" if they prefer to gift and lend within their own network. With more than 5,000 local Town groups and over 11 million members across the globe, the Freecycle Network offers a host of opportunities to give, receive, and save stuff from ending up in the landfill.
https://www.freecycle.org/

OTHER PLATFORMS.

Facebook Marketplace, Craigslist and NextDoor all contain categories of items listed for free.

LIBRARIES.

Libraries offer a lot more than books these days. Available offerings vary, and, if your town has a network of libraries, you may have to go to the main library or a different branch to borrow a specific item. Here's a list of the types of things and services many are offering:

Digital Media: ebooks, audiobooks, yes, but many also provide access to movies, music and TV shows.

Cultural Passes: free or discounted passes to local museums, theaters and attractions.

Educational Resources: adult education classes, language learning programs, supplemental lessons or tutoring for schoolkids, cooking instruction, home improvement how-to videos and more.

Meeting Spaces: free use of a room for hosting community meetings, study groups and more.

Maker Labs: sewing machines, embroidery machines, Cricut machines, craft supplies, 3D printers, laser cutters, engraving equipment, photography, 3D design equipment.

Technology and Gear: computers, tablets, internet access, VR headsets, gaming systems, cameras, projectors, and recording studio equipment.

Tools and Gadgets: drills, saws, sanders, musical instruments, fishing poles, gardening tools, stud finders, pressure washers and metal detectors.

Digitizing Tools: equipment for do-it-yourself digitization of photos, video, sound recordings, and other formats.

Kitchen Gear: cake pans, waffle irons, mixers, spiralizers, ice cream makers and knife sharpeners.

Leisure: sports equipment, games, puzzles, artwork, toys.

What does YOUR library offer?

See "75 Surprising Things You Can Borrow from the Library."

https://www.libbylife.com/2024-07-25-75-surprising-things-you-can-borrow-from-the-library

TIME BANKS

A time bank is a community where people exchange services, where time, not dollars, is the currency. Members offer services like tutoring, gardening, transportation, translating, photography. They earn a time bank credit equal to each hour they donate. These credits can then be cashed in to receive services from other members. It's an egalitarian system where each hour of service gets valued equally as any other. An hour of babysitting equals an hour of legal advice because all of our gifts matter. Time Banks can be found in most states in the U.S. and around the world.

To learn more, see https://timebanks.org/ and

https://hourworld.org/index.htm

Why join a time bank?

"The magic of time banking is the community building that is achieved. The best way to get started is to find an existing time bank as close to you as possible. Once you have joined go to any public events that they offer. That way you can meet other timebankers face to face."

David Cutter, Time Bank member

REPAIR CAFÉS

A Repair Café is a community event where skilled volunteers fix broken items for community members for free. In a world where it's often cheaper to buy new than repair, these events keep reusable things working and provide an antidote to that helpless feeling when we're surrounded by a bunch of broken stuff. Or maybe that's just me.

Types of repairs include mending clothing, fixing electronic gadgets, repairing broken furniture, fixing flat

> bicycle tires and broken toys. All services are provided
> free of charge and the café provides tools and basic
> materials. Some also offer space to exchange free items.
>
> Over 2,500 Repair Cafés around the world promote
> sustainability and community engagement.
> See https://www.repaircafe.org/en/

My Buy-Less-Amazon Hack

Much as I disparage the environmental scourge that Amazon is, I admit that there are times when it's the only option. If there's something I need that isn't available at my local shops, the nearby Target, the resale shops, or by direct order from the maker, then it's off to Amazon I go.

Case in point: the clip-on dimmable LED desk lamp with three color settings and a 27" gooseneck that's currently sitting in my cart. I've never seen one quite like it, but gosh darn if it wouldn't be the perfect utilitarian fix for my small office / craft room / guest room. Not exactly a life-or-death item, but I do get a charge out of finding Just The Right Thing. Maybe it's the Thrill of the Hunt, dating back to our hunting and foraging history. Or maybe I'm hopelessly trying to fill that inner void perpetuated by our consumer culture. Either way.

When the urge to acquire strikes, I let myself surf Amazon for options. Compare colors! Features! Eco-friendly packaging options! When I've selected the perfect candidate, I put it in my shopping cart and proceed to celebrating my savvy in ferreting out Just the Right One. I enjoy that dopamine hit for all it's worth.

Then I close the app. I tell myself that I'll wait to see if anything else pops up before ordering. Save emissions by ordering all the things at once. I can be eco-friendly even in my moments of overconsumption.

When the next urge hits, repeat. Wait. Repeat.

When the day comes that I hit upon something that I really do need—an odd-size battery, say, for a device that will NOT STOP CHIRPING—I pop it into my Amazon cart and survey the already-added items.

Nine times out of ten, the other items in the cart seem trivial in comparison to the Thing I Really Do Need. I find myself wondering: "*Why did I think I needed that?*" Or, "*I don't need that anymore because I've found an alternative.*" I winnow down to that one item, maybe two things, that I really really need. I kick the rest out of my cart. Which gives me another sort of thrill, not unlike kicking a bad-news boyfriend to the curb.

Waiting saves me money, reduces my carbon footprint, lessens my toll on Earth's resources—without giving up that dopamine hit of the successful hunt. It's a win-win of sorts.

I'm still torn about that desk lamp, though.

TWENTY-THREE
Crippling Ignorance

Good news! Climate change is over. Or never was, I'm not sure. Either way, the state of Florida has erased "climate change" from its statutes,[186] so we can all breathe a deep sigh of relief. The timing puzzles me, seeing as how that legislation was enacted as WVSN TV in Miami blasted this headline: "South Florida faces record spring temperatures; health officials urge caution." The heat index in several Florida cities ranged from 104° to 109° this past spring. But that's just too bad for workers because their state legislators enacted another law that prevents local governments from implementing any measures to protect outdoor workers from extreme heat.[187] Surely, they wouldn't have done that if global warming was a real thing, right?

In other news, the National Oceanic and Atmospheric Administration has predicted (as of this writing) an 85%

---

[186] Methinks a certain governor is taking notes from a former U.S. President who scrubbed all mentions of climate change from the White House website when in office. Copycat.
[187] Not just Florida. Texas outlawed ordinances in Austin and Dallas that mandated rest and water breaks for construction workers because of extreme heat—right after Texas endured three straight weeks of triple-digit temperatures. HB 2127 took effect Sept. 1, 2024.

chance of an above-normal hurricane season for 2024, which naturally prompted the busy Florida legislature to authorize showing videos that deny climate science to schoolchildren, kindergarten through 5th grade.

> *"Videos that compare climate activists to Nazis, portray solar and wind energy as environmentally ruinous and claim that current global heating is part of natural long-term cycles will be made available to young schoolchildren in Florida, after the state approved their use in its public school curriculum."*
>
> "Videos Denying Climate Science" in *The Guardian*

They are teaching kids that climate scientists are like Nazis?! As if climate researchers aren't already facing enough harassment.

Sure, let's teach the kids, the ones who will inherit the numerous eco-crises Florida faces, to vilify the people doing the essential climate research. Because, get this, Insurify, an insurance cost comparison platform, recently published "The 10 Worst Cities to Live in as Climate Change Progresses." Guess which state is home to 6 of the 10 worst cities, according to that report? Florida. Yes, taking every city in America into account, Insurify found that SIXTY PERCENT of the cities facing the worst of climate change are in Florida, noting that "the state is currently experiencing an insurance crisis driven by billion-dollar natural disasters."

Sounds bad. But maybe Insurify will revise their conclusions after they learn of the state's actions to eliminate climate change . . . with their legislative red pens.

I just can't wrap my mind around the heartlessness of lying to children on such a critical issue. I try to tell myself, maybe those legislators really do care, and, like overprotective, helicopter parents, they are trying to shield the little ones from the awful truth. Maybe they even want to believe it themselves. I know I wish I could wave a magic wand to make it all go away. Just say CLIMATE CHANGE IS A HOAX three times, click your heels and turn on Fox News for validation.

Well, no worries, if schoolchildren don't learn the truth in elementary school, we can always pin our hopes on the more rigorous studies of college. Except, wait a minute . . . I recently learned that oil companies like ExxonMobil, Shell and BP are some of the biggest funders of climate research at the most prestigious university research centers. MIT's Institute Energy Initiative is funded mostly by Shell, ExxonMobil and Chevron. ExxonMobil also funds Stanford's Global Climate, while UC Berkeley's Energy Biosciences Institute exists thanks to a $500 million deal with BP. The Center for Energy Economics at the University of Texas at Austin is endowed by Chevron and ExxonMobil.

That's the short list. And, not surprisingly, some major perks come from those multi-million dollar donations—like the oil company donors getting to decided which research projects get funded—and which don't. As a result, funding often gets directed to studies of far-off technologies, rather than ready-to-roll solutions.

Let me say that again because surely there's something amiss here. Fossil fuel companies fund climate research at major U.S. universities.

Make that make sense.

Never mind. You can't. It's as illogical as putting tobacco companies in charge of researching the harms of smoking. Which, oops, we also did, once upon a time.

Everything old (and horrible) is not new and still horrible. Why do we keep putting wolves in charge of the hen house?

It's a good thing these stories about dangerous heat and hurricanes are all just so much woke propaganda. Thankfully, we have the brave anti-woke crowd to tell us not to worry about all that. Protecting fossil fuel interests and making a few people stupid-rich is way more important than the well-being of all humanity and every other species on Earth. That's what they spend millions on advertorials to convince us, anyway.

And "they" say environmentalists are out of touch. Sheesh.

> *"I can say I don't believe in gravity, but if I step off a cliff I'm still going down . . . . When we decide not to react to the information that science provides us, it's a choice to increase our vulnerability rather than decrease it."*
>
> Katharine Hayhoe
> Climate scientist, Chief Scientist for The Nature Conservancy, and author of *Saving Us: A Climate Scientist's Case for Hope and Healing in a Divided World*

Keeping our young people in the dark about the world around them seems as stupid as neglecting to educate them about adulting basics like managing money, paying taxes, moderating social media intake, or even assembling furniture from Ikea. Ooops we're guilty of that, too.

What's next, pretending cheeseburgers aren't bad for our arteries? Ooops, we're pretty far down that path, too. Sheesh. Is there anything in our commercially sponsored world that we can believe?

Adulting is hard enough already in our ever-updating world. It seems I lose basic skills every day. Last month, I successfully paid the mortgage online, but, since the most recent website update, coupled with a forced password change, duplicating that accomplishment now seems like an impossible miracle.

But to intentionally cripple the intelligence of our

young people, who, after all, are inheriting the fossil-fueled messes we've made, well, it feels like Lucy pulling the football away just as Charlie Brown is about to kick it. If Charlie Brown's entire future depended on making that kick.

My kids are just starting their adulting years, and I feel for them. There's just so much more they need to know now—and so much more opportunity for public shaming on social media if you don't do it perfectly the first time. I thank God-Buddha-Allah and all the other deities that social media did not exist when I was an awkward teen . . . or even a stupid twenty-something. It's bad enough I have to replay those cringe moments in my own mind; I sure wouldn't want them broadcast out into the ethers.

I do know that my California-educated kids learned plenty about climate change and other environmental challenges in their public schools. I once mentioned climate change to my son, back when he was in eighth grade, and he cut me short. "I already know, Mom. And we're already depressed."

My heart broke. Again. I want nothing more than to protect my kids from the tragic truths of our time. But shielding them from scientific facts does not help prepare them for their futures.

Younger people are already stressed out, anxious and depressed. When you consider the price of housing, the rise in hate crimes against anyone considered "different," gig work providing no security or benefits, cringe culture ridiculing every mistake, and oh, their home planet being pronounced code blue? Those "we must take drastic action before 2030 or it's too late" warnings put a real damper on making plans for the future.

I grieve for every one of them. It's like arriving at the best party, and it's a real rager, and just when you start having fun, somebody yells, "COPS!" The sirens blare and everyone starts guzzling down the last of the hooch, swallowing fistfuls of pills, getting the last of the good stuff while they can. Everyone else runs, but you get arrested and pay the consequences for everyone else's fun.

That's effed up.

I just don't see how lying to them is going to ease their pain, any more than lying to myself about my weight helps me fit into my clothes.

Seeing the chaos at local school board meetings these days, is it even safe to attend to demand better environmental education for our kids? If you're a parent to school-age kids, I hope you have the gumption for it. Maybe wear some body armor.[188]

Can Sesame Street help? That show came to my emotional rescue countless times when I was a kid. Back then, I had to wait around for the show to air on TV. Nowadays, kids can access Sesame Street Workshops on many topics, from grief to understanding COVID-19, on demand. Er, whenever their parents allow screen time, that is. PBS has aired special episodes after major hurricanes, and the "Sesame Street Gets Through a Storm" episode on YouTube has more than 15 million views.

Sesame Workshop, the nonprofit producer of the

---

[188] And see the Activism Tips in the Aging Ragers chapter, too.

Sesame Street series, recently announced a partnership with the Save the Children nonprofit to "foster young children's climate resilience." They plan to create stories designed to help kids and their families both understand why weather extremes like hurricanes, droughts and floods are happening, but also how to prepare for them.

Call me crazy (many do) but preparing kids for what's coming makes a lot more sense than sweeping it all under the climate-change-is-a-hoax carpet. Let's not make this one more shell game, hiding hard truths, and costing them precious opportunities to do SOMETHING. We're not sparing them anything except even greater anxiety and depression when they learn the truth.

I know kids have a zillion more choices for entertainment than I had at their age, so it's impossible to say how many kids this programming will reach. Still, Sesame Street is said to be the largest "informal educator" of young children, reaching some 154 million kids in 150 countries around the world, delivered in 17 languages, so I am hoping that the number is SCADS.

GET ON BOARD

As much as I have supported my children's education, the thought of sitting through a school board meeting cranks up my excuse-making machine to hyperdrive. I pray you possess the fortitude, if you have school-age kids today. The squeaky wheels do get the grease, and the squeakers at your local school board meeting may not represent the best interests of your family.

Check out this "5 Very Good Reasons to Attend a School Board Meeting" article from Parenting.com for tips on what to expect and how to get involved. Maybe if I'd read it before my kids graduated, I'd have summoned the will to attend one myself.

https://www.parents.com/parenting/better-parenting/very-good-reasons-to-attend-a-school-board-meeting/

GET OUTSIDE

The more time we spend in Nature, the more we care about it—and the more likely we are to do something to protect it. That's not just me pontificating, not this time. A 2021 study published in the International Journal of Environmental Research and Public Health backs me up.

"Substantial evidence from observational and intervention studies indicates that overall time spent in nature is associated with increased perceived value for and connection to nature and, subsequently, greater pro-environmental attitudes and behaviors."

Time in Nature offers a host of other benefits, too,

including:

* Reduced stress

* Reduced risk of Type II diabetes
* Improved concentration
* Improved creative problem-solving skills
* Reduced inflammation
* Improved mood

Sign me up for all of that. Gosh, with so many benefits, I wonder why we don't ever hear NATURE advertised on TV?

And it's affordable. While a walk in the woods is top-notch therapy, even a mindful stroll down the block, stopping to study the fallen leaves, the snails sliming across the sidewalk, or the sparrows singing overhead, contributes to our well-being.

With all the pressures kids face today, time in nature could just be the therapy they need right now. Heck, it could be the therapy we ALL need right now.

GET OUT OF SCHOOL

I volunteered to chaperone many a field trip when my kids were younger, mainly to museums and a few amusement parks. I recall only three of them spent outdoors—even here in the mild weather of Southern California. On one of those outdoor trips, the class witnessed a wild coyote snatch an unsuspecting squirrel into its jaws, then carry it off into the meadow for lunch.

The entire 5th grade class, once boisterous to be out of the classroom, fell silent. More impactful than any episode of *Wild Kingdom* I've seen.

Need ideas for outdoor education? Check out "**Beyond a Book**" for adventure-linked education projects that go beyond the science lessons to discover, explore and learn to protect the diversity of life on Planet Earth. Beyond a Book offers lesson plans, videos, field trips, and more to transform real-world experiences into lessons with limericks, letter writing, math problems, art projects and other creative projects.

https://www.beyondabook.org/

TWENTY-FOUR
Stench: When Eco-Intentions Turn Rancid

Stabbing with the sharpest knife in the kitchen, I am determined to get this puke-colored slab out of my eco-friendly Pyrex bowl. My most violent stabs make only the slightest of dents, about as effective as digging a well with a toothpick. But I soldier on in Missouri mule fashion because I've wrapped my entire identity around my Earth-loving ideals, which means my very survival depends on rescuing this Pyrex. Mid-strike, I glimpse a dog walker outside my kitchen window, staring, aghast. I drop my weapon long enough to draw the shades. I can't risk another alert on the Scary Neighbor Watch message board.

From behind the blackout shade, I'm back at the whacking, giving it my all. If this gunk defrosts before I can chuck it, releasing the stench of it into the open air, I may not live to tell the tale. It's Trash Day, so I have just minutes to rid this retch from our lives. The big truck that makes our detritus disappear—or so it seems—should be here any minute. Don't get me started on the

Landfills of Despair where nothing really composts, not even the 100% biological materials. I'm already mad enough, and I'm wielding a deadly dagger.

Freezing this barf-colored glop seemed like an obvious, lifesaving decision back when its vulture vomit stench nearly inspired a call to the EMTs. Now it seems like yet another failure of living up to my Earth-Loving Lifestyle imperative. A child of the eco-era of the first Earth Day, the newly-minted Environmental Protection Agency, and the Whole Earth catalog, I've been intent on living eco-lightly ever since I first masking taped my bedroom light switches in the OFF position, back in the Energy Crisis of the 1970s. My parents were not amused, by the way. No pats on the back for living the lessons learned in my ecology class; just a thoughtless "I pay the light bill, I decide ..." sort of response. Was that "no TV for a week" punishment supposed to be ironic?

Infuriated by their lack of concern for the dwindling resources of our spaceship planet Earth, I determined then and there that I'd do whatever I could to get back at them. I mean, I'd do what I could to make a difference. I wrapped my very identity in the spring green cloak of eco-sanctimony, and I've never looked back.

But how did my good intentions lead to this moment of mad-stabbing at frozen sludge? Yet again my do-gooder dreams have led me down a well-worn and well-meaning path that suddenly took a turn for the ludicrous.

Don't buy new when you can reuse.

Jars. I wanted empty baby food jars to store the watercolor paints I'd mixed for my zine. How I got started on a "comic book," so to speak, at my, *ahem, advancing* age is yet another story. Suffice it to say there's a big section of my hippocampus that's still convinced I'm youthful, ageless, and hot. And I am not talking about menopause there. But I digress.

I have a myriad of empty glass vessels in my kitchen, given my inability to toss those *perfectly good* Mason jars that once held marinara or dill pickles. But they are all too cavernous for the smattering of watercolors I want to mix up. Twenty years ago, I could not have imagined life without a steady stream of baby food jars, birthing two babies just sixteen months apart. But my stash of tiny jars vanished long ago, like so many sock mates in the dryer.

Buying new jars of baby food, just to toss out the contents, was not an eco-friendly option. Think of all the water, the backbreaking farm labor, and the natural resources that went into making that food!

So, I posted a request on my local Buy Nothing group for empty baby food jars. I was dumbfounded to receive only one reply on Buy Nothing: a mom offered me a bunch of baby food jars. With. Food. Still. In. Them. Who says the Universe doesn't have a twisted sense of humor?

Turns out, her little cherub, an early adopter of veganism, perhaps, refused to consume the meaty contents, so the jars were mine for the asking. Jars full of food. I smacked myself on the forehead and groaned.

Then a stroke of genius landed, Divine Providence at work, and I hit upon the idea of feeding the 100% beef baby food to my carnivorous cats. Voilà! No wasted food. Google says this baby food brand is highly recommended for cats with "sensitive tummies." Given the volume of cat vomit we see in this house, I readily assumed our cats to be of the sensitive tummy variety.

Eager to get painting, I emptied the baby food from all twelve jars into this Pyrex, with plans to dole out the deliciousness to the fawning felines sparingly. It's 100% beef, after all! I imagined how my cats would worship me, and tomcats down the block would howl in envy, just to catch a whiff of the beefy aroma.

The emptied and cleaned jars worked perfectly for storing my small batches of watercolors. Basking in the smug glow of eco-sanctimony, I braced myself for the onslaught of adoration when I fed the baby food to my furred family members.

As a long-time cat custodian, you'd think I'd know better.

All three cats—who never agree on anything—were united on this critical point: shun this puke-colored offering like vegans at a pig roast. One cat pouted, giving me a hurt look as if I could never be trusted again. The second one sulked away in disgust. The third one barfed into my favorite slipper.

Seething, I flung the Pyrex into the back of the fridge, vowing to feed those cats only kale for a week. *See how they like that!*

Days later, a foul odor assaulted me when I opened the fridge door. No worries, I've learned to slam it quickly in such situations—well before my husband can suggest that "the fridge needs cleaning out!"

Two more days later, I couldn't slam fast enough, and the putrid stench threatened my very state of consciousness. In a coordinated move that rivaled

Simone Biles on the beam, I flung open the fridge door, grabbed that wretchedness, stuffed it into the freezer, and slammed shut both doors with such aplomb that I managed to avoid revisiting the contents of my stomach.

The reek vanished in just a couple of hours. I breathed a self-satisfied sigh of accomplishment. In a not-so-eco-friendly move, I opened and closed the freezer door several times, relishing the odorless air.

Have I mentioned how much I love my freezer? It's been there for me in times of stress like few appliances have. Want to get gum off your shoe? Toss it in! Got bugs in the rice bag again? Toss it in! Got a beeping toy that WILL NOT STOP in the middle of the night? Toss it in!

But this life-saving hack was my favorite so far.

Until now. As I'm stabbing with futility, desperate to defrost this rotting sludge enough to free my Pyrex, while also dreading the stink of the defrosted goo, I realize that COVID-19 taught me something useful.

MASKS SAVE LIVES

I grab a handful of masks from the stash and don them all. I return to my work with the smug satisfaction of an old dog who can, in fact, learn new tricks. Just a hundred more whacks or so and this will all be over.

I deliver the glop to the trash bin moments before the trash truck arrives. The driver pays no mind to my celebratory gyrating—or at least he politely pretends not to notice. I do hope my shenanigans eased the monotony of driving his regular route day after day.

Back in the kitchen, I check the clock. Only 9 a.m.! I'm

as exhausted as that time I got lost hiking in the weeds in my yard.[189] When I think of all the time I've devoted to obtaining a few small jars . . . well, maybe it's time to re-examine my self-identification as the consummate Green Goddess. Give myself a break. Do the best I can, when I can, however I can, without losing my mind over some watercolor paint containers.

Either that or up my blood pressure medication. I doubt the CEOs of plastic manufacturers, the oil company tycoons, or Rupert Murdoch lose any sleep over the destruction and chaos they wreak. I watched *Succession*. And they get paid bazillions. Why should I appoint myself the sole, unpaid receptacle of all our collective eco-guilt?

Savoring a cup of fair-trade, shade-grown coffee, I smile to imagine myself just letting eco-guilt go. It's an odd sensation, but I'm enjoying the exercise. So freeing. I could get used to this.

Then, I smell something. Before I can blame the cats, I espy my old gym bag full of sweaty attire that's been sitting for who knows how long. Somebody should do something about that.

---

[189] Note that my yard is "Los Angeles-sized," lacking the vast acreage found in the Midwest neighborhood I grew up in.

TWENTY-FIVE
Ranting In The Street

A woman was ranting out in the middle of my street this morning.

You may be surprised to learn that it was not me.

I sure was.

I spotted her at the bottom of our hill, waving her arms and hollering, as soon as I began my jaunty descent. The descent, I might add, is always jaunty. Funny how the *ascent* never is.

Late for my appointment, per usual, I was half jogging while straining to make out her complaints, all the while wondering... how is this NOT ME? I've been feeling consumed with some furious thoughts lately. In fact, the volume on my own inner rants has increased to the point I'm wondering if there's a rave happening in there.

Not that I've ever been to a rave. I can't stand loud music anymore, and I'm in bed before they start. But I was young once, and I have the tinnitus to prove it.

The cacophony of my inner ranting booms so thunderously now that I often wonder if anyone else can hear it. Is this person in the street a projection of my own inner turmoil?

When I got closer, I realized her tirade concerned a failed love affair. I breathed such a deep sigh of relief that I nearly asphyxiated myself. Not my rant party. Even when I'm fuming over a million injustices, the stability of my own family situation is the bedrock that keeps me grounded. And for this, I do give thanks.

Still, the rant rave in my mind rages on, full of judgements of Other People, the ones hellbent on desecrating all that I hold dear. Their utter disregard for the environment and climate that sustains Life as We Know It makes me as mad as a customer service rep treating me like I'm already senile. The denial of life-saving health care for women, the disregard of indigenous people, and the despising of LGBTQ and disabled persons make we wish I could just burn the Internet down.

In other words, I'm destroying my own inner peace every time I judge what The Other People are doing.

WHY do I do this to myself? Does my tranquility mean so little that I'd give it up so easily?

I know I'm not alone in my emotional distress. We've all experienced some level of upheaval in the past couple of years, and I'm not just talking about Elon Musk gutting Twitter. Many, many Americans say their mental and emotional well-being have suffered since Covid descended, since inflation flared, since the political rancor divided us, since job security went the way of bell bottoms. Far too many young people worry so much about climate change and their body images on Instagram that they contemplate suicide.

It's not an easy time to be alive. Not if you're paying

(too much) attention to the news. Or experiencing homelessness, joblessness, loneliness, ill health, loss of loved ones to Covid...

Much as I kid about my mental well-being, I do prioritize caring for it. ~~Everyday~~ ~~Most days~~ On good days, I walk myself around the neighborhood and sit in meditation long enough to reconnect with my own Inner Peace. Revel in the utter ecstasy of No Thought. When I'm lucky enough to shut the monsters down, that is. Some days are better than others.

And when that doesn't work, I book myself an appointment at the Rage Room, a magical place where they give you stuff to smash and a hammer. I once thought that sounded wasteful until a few Supreme Court decisions dropped. Now it sounds like a gift from on high. I get to vent my rage and someone else cleans up the mess? I am IN.

I devoted hours to compiling a list of simple and effective healing therapies for my book, *Love Earth Now*. It's time for me to give it a review and find more ways of investing in my emotional well-being which don't require smashing stuff. I'll reprise it here, for us all to peruse. What's your go-to therapy for venting the hard emotions?

Today, I'm taking a break from my own ranting by reading some Mary Oliver poetry. Doesn't sound like much but it does clear my mind, and that's enough to keep me from ranting in the middle of my street. For today, anyway.

THERAPIES FOR VENTING AND THRIVING

Where do you turn when the steady stream of bad news seems too much to bear? What's on your list of go-to therapies?

Do you hit the hiking trail, journal, rant on social media, pray, cry, stop and smell the roses?

Flick a fidget spinner, listen to music, meditate, take a bubble bath, roast marshmallows over a blazing fire?

Drive the backroads, send a thank-you note, volunteer to rock newborns, teach a class, paint the bathroom?

Bake cookies, knit baby booties, pray the rosary, retreat to the man cave, sing in the shower, eat mac and cheese?

Stretch, nap with the dog, light a candle, throw darts, binge-watch Netflix, camp out in the backyard?

Go to church/temple/mosque, cuddle the cat, hit the driving range, hula hoop, scrub the shower?

Pore over old photo albums, donate to charity, plant seeds, read a romance novel, dance like nobody's watching?

Walk the dog, do Sudoku, clean the house, go to a museum, navel gaze, walk the beach, do needlepoint?

Go to the gym, climb a tree, read a spiritual text, listen to a podcast, soak in the tub, take a moonlit stroll?

Make sandwiches at a shelter, foster kittens, do tai chi, watch funny videos, make a gratitude list?

Call a friend, pull weeds, leave a secret gift on your neighbor's porch, clean out the junk drawer?

Burn sage, paint your toenails, get frisky, chew gum, kick something, talk to a tree, do a jigsaw puzzle?

Blow up a balloon and pop it, hit the hot tub, do a crossword, call Mom, write a song or poem?

Color outside the lines, play an instrument, wear a funny hat to the grocery store, drink sun tea, write a positive review for a friend?

Make a prayer bundle, call your representatives, gnaw furiously on a baguette, get a tarot reading?

Chat with your therapist, make a collage, take food to the community fridge, can tomatoes?

Lie on a blanket and cloud watch, scale back outlandish expectations, give yourself permission to do nothing?

Consider making your own list of go-to therapies, including the tried-and-true favorites as well as some you've never considered before, something that sounds FUN. Don't we need more fun in our lives right now?

"Rather than letting our negativity get the better of us, we could acknowledge that right now we feel like a piece of shit and not be squeamish about taking a good look."

Pema Chödrön
Tibetan Buddhist teacher, nun and author

TWENTY-SIX
Environ-Mental: Reinventing Our Brand

"She's not even a vegan!" This comment in a review of my book, *Love Earth Now*, makes me smack myself with a copy of said book. "This is why they despise us," I seethe, each "s" hissing like an asp on acid. Much as I deplore all the judgi-ness in our divided society, it's the "not-green-enough" accusations that stab me in my Earth-loving heart. Since when did caring about the health of our natural world devolve into eco-purity testing? Only people who live off the grid, wear hemp sackcloth, and live in a yurt can comment on the destruction of our environmental support systems? Namaste, tree huggers.

True confession time: I don't call myself an environmentalist. Not anymore. Sure, I wore that label loud and proud, back in the Days of Innocence (i.e., before the Internet). Nowadays, it's just too heavy a burden to bear, considering all the baggage that certain uninformed-but-outspoken bashers[190] have heaped onto

---

[190] People who say things like wildfires are being caused by lasers or space solar generators. Stuff like that.

it. Figuring out what on Earth any of us can do to tackle even one of the monumental challenges we face is hard enough. Imagine trying to do it with a scarlet letter pinned to your sackcloth shirt.

How have "we"—meaning us renegades who want our children to enjoy breathable air and drinkable water—lost control of the messaging? Something has been lost in translation since those (in retrospect) glory days of a conservative U.S. President, George H. W. Bush, insisting we leave this Earth in a better state than we found it in. He'd be excoriated on conservative media for spewing such woke nonsense these days. WTF happened?

"[W]hy does environmentalism provoke so much annoyance, alienation, and downright rage?"

Jenny Price
Author, *Stop Saving the Planet*

I'd like to blame it all on the oil companies that invested so much in studying climate science back in the 1970s and 1980s and then buried the truth. And that's not a stretch. Thanks to some leaked internal memos, we now know that Exxon and Mobil both knew the dangers of climate change as early as the mid-1970s—and then did everything they could to pretend they didn't. Like toddlers with crumbs all over their faces saying they didn't touch the cookie jar. Only we are just now finding the cookie crumbs.

> *"In 2015, investigative journalists from the Los Angeles Times and The Guardian . . . found dozens of internal documents from Exxon and ExxonMobil scientists that clearly detailed how fossil fuel products contributed to the global climate crisis—which could have "dramatic environmental effects before the year 2050."*
>
> Carly Cassella
> "ExxonMobil Predicted the Climate Crisis"

Back in the '70s, Exxon had fully funded and staffed a research lab, intending it to be the "Bell Labs[191]" of cutting-edge energy research. Exxon invested in top scientists exploring everything from the greenhouse effect to renewable energy sources, intending to be an Energy Company, not limited to just fossil fuels.

But one sad day in the early 1980s—and if I knew the date, I'd declare it a day of mourning—the "holistic-energy-company" image went down in flames. Henceforth, the company doubled down on promoting fossil fuels, shuttering much of the alternative energy research, and sowed doubt on the climate science that their own scientists had carried out. That's not a case of one hand not talking to the other. It's more like one hand denying the other hand exists.

---

[191] "Bell Labs" referring to the cutting-edge research laboratory founded by Alexander Graham Bell with the money he won for inventing the telephone. Dubbed "Bell Labs," it's where numerous ground-breaking technologies were first invented—and still are today.

MAKE THIS MAKE SENSE. Exxon funded the climate science in their own labs, then published weekly "op-<u>ads</u>"[192] in the *New York Times* from 1972 to the 2000s to convince us all that the science conducted by their own scientists wasn't "settled" or sufficiently certain.

After a rigorous, academic review of in-house documents, publications and advertorials by ExxonMobil over 40 years, two *New York Times* op-ed contributors determined that ExxonMobil did indeed know better.

> *"Exxon Mobil misled the public about the state of climate science and its implications. Available documents show a systematic, quantifiable discrepancy between what Exxon Mobil's scientists and executives discussed about climate change in private and in academic circles, and what it presented to the general public."*
> Geoffrey Supran and Naomi Oreskes
> *New York Times* op-ed

The predictions of the warming trends by their own scientists proved remarkably accurate, too. Harvard researchers found that between 63 and 93% of Exxon's global warming projections produced between 1977 and 2003 were accurate. But somehow 81% of their climate change "advertorials" expressed doubt about the science

---

[192] That's not a typo. Op-eds give opinions. These "op-<u>ads</u>" or "advertorials" were written to convince the American people that oil companies were trustworthy and that climate science wasn't.

in one way or another. *What cookies, Mom?*

No wonder people are confused. Skeptical. In denial. Because we're not talking about the dubiously-sourced *Enquirer* here. These were published in the *New York Times,* week after week, for decades.

Or maybe I'll blame the politicians who play their own shell games, eliminating climate change from state laws and repealing energy conservation targets even as their citizens suffer record-breaking heat. While on that darned other hand, they are funding "resilience projects" to the tune of hundreds of millions of dollars.

"Today, Governor Ron DeSantis signed into law Senate Bill 1954. . . . The new program will enhance our efforts to protect our inland waterways, coastlines, shores and coral reefs, which serve as invaluable natural defenses against sea level rise."[193]

Office of Governor Ron DeSantis

We really need to be checking both hands before we go blindly believing what we read on the Internet, except, whoops, most of the time we don't do any fact-checking at all.[194]

Or how about federal legislators bragging on new

---

[193] Sea level rise? Gosh, that's curious. What could be causing that, if climate change has been eradicated in Florida??
[194] I rely on sites like mediabiasfactcheck.com and FactCheck.org to get a sense of the truth (or the best we can get these days) before starting a new rant.

"green" projects which were funded by the Inflation Reduction Act—which they voted against? Case in point: Lauren Boebert claimed credit for securing over $50 million in funding for the South Bridge project in Glenwood, Colorado. She voted against the Infrastructure Investment and Jobs Act of 2021 that authorized the funding.

Or state legislators enacting crippling restrictions on renewable energy projects, even though they represent a fast-growing job sector? In Texas, where reliability of the energy grid is a concern, especially since the power crisis of 2021,[195] solar energy projects are ramping up, creating scads of new, well-paying jobs.[196] So why is the Texas legislature still passing anti-renewable energy laws?

How do they even keep their stories straight? *Okay, I'll be FOR this at the PR event on Tuesday, because it's popular with my constituents, but I'll rail against it on social media on Wednesday because it's "woke"?* Yet another reason I could never be a politician, and I'm not just talking about, *shall we say*, the "experimentations" of my youth. I have trouble just remembering whether I said I'm definitely "on a diet" or "diets are the worst way to lose weight." On any given day, it could go either way.

But back to our conflicted politicians. I get that they are pandering to the anti-woke crowd, a not insubstantial sector of voters, whose sole agenda seems to be ridiculing

---

[195] My sister lives in Austin, and she ended up in the hospital after four days in below-freezing indoor temperatures without heat or running water

[196] The U.S. Department of Energy reported 114,000 clean energy jobs were added nationally in 2002. Of these, California, West Virginia and Texas experienced the most rapid job growth in the clean energy sector.

anyone or anything that didn't represent The Norm circa 1950. Back when life was swell. If you were white[197], male[198], heterosexual, cisgendered, cognitively intact,"[199] and middle-class (or better), that is.

Those olden times weren't so swell for Earthkind[200] either. Back when unregulated use of pesticides like DDT nearly caused the extinction of songbirds and bald eagles.

"On the mornings that had once throbbed with the dawn chorus of robins, catbirds, doves, jays, wrens, and scores of other bird voices, there was now no sound; only silence lay over the fields and woods and marsh."

Silent Spring by Rachel Carson

Back when factories dumped untreated waste directly into rivers—so much oil-soaked debris had been discharged into the Cuyahoga River in Ohio that it caught fire in 1969. Back when coal-fired power plants and

---

[197] I have a Black friend who was chosen to attend the nearby and newly desegrated elementary school in Little Rock but rode the bus to the colored school an hour away because her family feared the violence. Not so swell times for her.
[198] My mom was a teenager in the '50s and she could not wait to leave that decade behind. She embraced every new development that relieved women of homemaking drudgery, from TV dinners to birth control.
[199] My dad had a sister who was deemed "retarded" at a young age and locked up in a county facility because that's "how it was" back then. The family was told not to visit because it would be too "traumatic." Life wasn't swell for her either.
[200] "Earthkind" is my word encompassing both humankind and also all living species.

vehicles lacked any emission controls, producing killer smog. It was so thick over 5 days in London, the "Great Smog of 1952," that more than 4,000 died and 100,000 suffered from respiratory illnesses. Read that again: after just FIVE DAYS.

Bad news, folks. No matter how the U.S. Supreme Court may enable the wonkers,[201] we aren't going back to a time when we treated Earth like a dump, back when women and people of color had fewer rights than rocks. Project 2025 notwithstanding, that's what I must tell myself if I am to make it through the day sober.

But all that said. . . .[202], I have to acknowledge that our eco-friendly messaging sucks. We make it sound like caring for Earth is dreary and dour, all about scrimping and self-sacrificing, with "achievement awards" for those who have flogged themselves in the name of Earth. Demanding everyone else give up the good things in life, too, so we can all be dreary and dour together.

That's going to be as popular as McDonald's pushing bean sprouts instead of fries. As un-American as suggesting that money can't buy happiness. Or that super-sizing for a few cents more (and hundreds more empty calories) isn't exactly a savvy savings plan.

I wish I knew how to fix the branding hole we're in, but marketing just isn't my area. Unless we can convince people that caring for Earth reduces wrinkles, pays off

---

[201] This is the politest term I know to refer to people who want to undo such progressive notions as allowing anyone to marry, use birth control, practice their own religion (or not) and have access to lifesaving health care.
[202] The number of detours taken to get to my point here rivals my best efforts. Google Maps, I am not.

your mortgage, tastes as good as cheeseburgers, and results in weight loss faster than Ozempic, I'm hard-pressed to know how to sell caring for our life-sustaining environment. If the heat deaths, the superstorms, the mega wildfires, the bleached coral, the rising seas don't trigger any calls to action, what will?

Maybe what's needed most is for us who care so very much to just stop pointing our sanctimonious fingers. Tempting as it is to blast out those virtue-signaling "I just sewed my own sackcloth yurt!" posts, how about we take a moment and consider the intended impact of the post. Is it to inspire others to action? Or to secure our spots in our own silos?

Because the last time I got inspired to action by reading yet another inflammatory, "you just don't get it, but I do!" post was never.[203] They drive me straight to defensiveness, desperate to prove the person who posted it an idiot, without giving a moment's reflection to my own position. Maybe that's just me, but, judging from the growing divisiveness across the United States (and beyond)... I suspect a lot of people react as I do.

Once upon a time, being friends with someone whose political or religious beliefs differed from mine was not only possible but happened a lot. Me and my ... *let's go with* ... "quirky" thoughts and beliefs, well, I don't find many who think exactly like I do, even amongst kindred spirits. I would spend a lot of time alone if I didn't make space in the sandbox for some kids who built different sorts of castles than I did.

---

[203] I do realize that I just did the judge-y thing I am complaining about. This diatribe is intended for me as much for me as anyone.

ENVIRON-MENTAL: REINVENTING OUR BRAND

Of course, I didn't always know, back in the Before Times, exactly where friends stood on critical issues. We could meet for coffee, say, and I did not review their socials[204] before deciding what topics to avoid—or whether to unfriend them forever. Heck, "un-friending" wasn't even a word back then.

May sound crazy, kids, but it's true. I was friends with many a staunch conservative, and I didn't lie awake at night fuming over their idiocy, either. Nope! We just talked about good movies, favorite restaurants and whether pineapple belongs on a pizza. Or some such drivel,[205] depending on how much alcohol was consumed.[206]

How did we get here, hurling vitriol from our respective ideological silos? Nearly 80% of American now report having few or no friends "across the aisle," according to a 2020 Pew Research study.

THIS IS WHAT KEEPS ME UP AT NIGHT.

More than anything. Yes, we have huge problems to solve and very little time to accomplish them. But there _are_ solutions in many cases,[207] and research for more effective and affordable solutions goes on. But we will never be able to implement any of them, not on a

---

[204] BECAUSE THERE WEREN'T ANY.
[205] Did you know that licking a stamp burns 1/10 of a calorie? But the U.S. Postal Service is phasing them out in favor of the self-adhesives. So, if you're overweight like me, blame the Postal Service.
[206] Joke memes did exist back then, they just arrived via an antiquated machine we called a "fax" which would spit out pages you didn't print yourself. They'd just show up like spam, only you had to buy the paper and toner.
[207] Bill Gates offers a thorough, issue-by-issue analysis of viable solutions (and where more research is needed) in his book, How to Avoid a Climate Disaster.

widespread scale, anyway, if we can't even talk to each other.

Reaching across the aisle does seem easier when there's a specific, local issue at hand—rather than cross-ranting about issues in the abstract. Look at the opposition in West Virginia to the Mountain Valley Pipeline. My friend in Greenville reports that concerned conservatives and liberals alike worked side-by-side to protest the pipeline crossing their lands because it was personal to each of them. When their multigenerational family farm is decimated, their well water muddied, or their municipality's water supply contaminated, concerned West Virginians didn't check political affiliations before taking action; they just got to work, doing what they can to preserve their ways of life.

Could we possibly learn from their example? Because we may have to take this community-by-community, now that the federal environmental agencies have been hamstrung by recent U.S. Supreme Court decisions. Let's hope and pray that enough people in the path of pipelines, in the backyards of petrochemical plants and downwind of oil fields can muster the will (and the lawyers) to challenge the polluting-is-essential-for-progress status quo.[208]

---

[208] Don't get me started on another "HOW IS THIS FAIR?" rant. The environmental justice movement focuses on pollution that disproportionately impacts poor and minority communities. But why are we still having these conversations when the harmful impacts from drilling, processing and burning fossil fuels are well documented? Why is it up to neighbors to stop it? Makes as much sense as saying it's okay to build a porn shop next to an elementary school unless neighbors take effective legal action because zoning laws are anti-progress. END OF SIDE RANT.

I'd be more encouraged if the Mountain Valley Pipeline had been stopped—but it's now operational. Or if I hadn't read *An Unreasonable Woman* by author Diane Wilson, a fourth-generation shrimp-boat captain and mother of five. The book chronicles her valiant efforts to stop petrochemical plants from dumping lethal poisons into the bays along her beloved Texas Gulf Coast—and all the ways she was vilified, scorned and threatened by her neighbors and the multibillion-dollar corporate polluter for her efforts.

Let me say that again. A mother and fourth-generation resident living near a petrochemical plant polluting the bay where she operated a shrimping business, dared to demand that the illegal polluting end, for the health of her children, for the health of the entire shrimping industry. For this, she received death threats. [209]

The good news is that in 2019 a federal judge found the polluter a serial violator with more than 1800 violations of clean water laws. The resulting settlement required the polluter to pay $50 million, earmarked for local environmental projects. Too bad it took something like three decades of activism—including the author's jail time and grievous emotional tolls—to achieve it. How many of us are willing to go to such extremes?

---

[209] Reminds me of a certain scientist who was demonized and threatened for suggesting Americans wear masks in the midst of a global pandemic. But don't get me started down that rabbit hole.

> *[Wilson's book] is the rare, clear, moving voice of a working-class woman goaded into action against the greatest massed forces in the world today: globalized corporate greed backed by government power.*
>
> *Going up against all that can make you feel slightly outmanned and outgunned. But Diane Wilson has discovered a weapon I believe is the greatest strength of many women: pure, cussed stubbornness. She is an unreasonable woman. God bless her. Unreasonable women may yet save the world.*
>
> Molly Ivans

Columnist, author, political commentator

Would Diane Wilson have stopped before she started her activism if she'd known how difficult it would be? Given her "pure, cussed stubbornness," I suspect not.

Here's to unreasonable women. And cussed stubborn Earth lovers everywhere.

CUSSED STUBBORN EARTH LOVERS. It's not "weight loss for everyone!", but it's the best brand idea I've come up with, given my poor marketing skills. What have you got? What's YOUR suggestion for our new and improved Earth lover brand?

A FEW UNREASONABLE PEOPLE

If you're looking for a few unreasonable people to inspire your own community-based action, consider these examples.

Northwest (Ontario) Climate Gathering

"THROW A PARTY." That's the advice of Heather McLeod, host of the "Something Different Comes This Way" podcast and co-founder of Northwest (Ontario) Climate Gathering. Who wants to attend yet another boring old meeting, when there's a party going on?

The Northwest Climate Gathering is "an intergenerational group working to connect communities in Northwestern Ontario to take positive, solution-oriented climate action." Designed to re-energize the climate carers[210] in their area, the Gathering employs that sneaky tactic called "FUN" to bring people together to discuss, yes, fears, but also stories of hope, ideas for increased resilience, and opportunities for community-based action.

But how to find your people? Heather McLeod recommends, "Think of two people, they could be one of your kids, your neighbors or your friends. Invite them over for tea to discuss what we're worried about, what

---

[210] That's a new term I just made up because who's going to be anti-caring? Don't answer that.

we're hopeful for, and what could we do that would make us feel like we are making a difference."

Toss out the unrealistic "go big or go home" expectations and allow your group of carers to grow organically, starting with just two or three folks. We put enough pressure on ourselves just to portray our lives as perfect without heaping on performance anxiety about our activism.

https://www.nwclimategathering.ca/

Plastic-Free Aberporth

Come meet the 1,100 residents of a small Welsh village, Aberporth, who have reduced their consumption of disposable plastics to ZERO. My heroes.

Local filmmaker Gail Tudor spearheaded the project after a tour around the British coastline where she witnessed far too much plastic pollution. She started a Facebook group, to inform and engage her neighbors, and the movement took off from there. Check out their group page and website:
https://www.facebook.com/plasticfreeAberporth/

https://www.iberdrola.com/sustainability/living-without-plastics-aberporth

Denver Urban Gardens

Denver Urban Gardens started as a grassroots movement in the late 1970s when some gardeners created a space for a group of local Hmong women to grow their own food. Working collaboratively, they transformed a vacant parking lot into what is now the Pecos Community Garden. Word of that first project spread, and inspired many more community gardens as the group helped other neighbors do the same.

Each year, the urban gardens across Denver distribute 625,000 pounds of food to community members. There are now more than 200 community and school-based gardens and 24 food forests across seven counties in metro Denver.

"Engage with Denver Urban Gardens to grow your own organic produce, become part of a wonderful community and do your part in combating the climate crises by healing the soil, enhancing biodiversity and reducing the heat island effect. Join a garden, come to a class, participate in the Grow a Garden program or volunteer!"

Linda Appel Lipsins, Executive Director

https://dug.org/

TWENTY-SEVEN
Guerilla Foraging

Uh-oh, I just got an email from Nextdoor, subject line: "BEWARE THIS WOMAN." My hand hovers over the keyboard as I pause to wonder if I dare read this email. Have I finally been caught?

Just moments ago, I was gushing over my plunder, the bounty of another urban foraging escapade. I was sure I'd made it home without being video shamed.[211] But now I wonder...

Thanks to a complicated series of bribes and threats, I entice myself into a brisk walk every morning, an essential part of my daily "How to Survive Another Day in Eco-Insanity" routine. And I say, "*every* morning" with a wink and a nod because some days are more fraught with excuses than others. "Too hot" has been the leading cause of my commitment demise lately, so I've been coercing myself out earlier and earlier—to a time I'd call "still night" on the dark mornings of winter.

I always trek the same blocks around my

---

[211] This is what I tell myself every time I walk through my front door. It's a delirious dream, considering the whole point of social media seems to be: who can we mortify today?

GUERILLA FORAGING

neighborhood, so many times now that I've memorized every hedge, picket fence and pee-stained fire hydrant along the way. My overactive Gemini mind usually detests—eschews at all costs!—any routine smacking of "same old, same old," so it was time to up my ante, if I wanted to keep up the walking habit.

My daughter gave me a gel plate for my birthday, and I've gone mad with making prints of flowers, leaves and seeds. So fun and so easy! Slap on some paint, stick on some leaves, press on some paper, and voilà! Art in 60 seconds and so little patience required. Just my size.

I dearly love creating art from things found in nature, as a way of reminding myself of the gifts we've been given. The gifts I mean to protect. And the process of creating art puts me in that beloved flow state, where I'm blissfully oblivious to the devious machinations of the Supreme Court.

As with any addiction, though—and I am familiar with several—I soon needed more, more, more. And I knew just where I could get them. My neighbors have planted such a variety of vegetation, so many fascinating leaflets, weed stalks and tiny blooms, that the possibilities for printing seem endless. And all of them, by my way of calculating, FREE.[212]

Suddenly, my daily slogs became a guise for my stealthy foraging operation. I'm now a spy, a thief, a double agent. For thirty minutes a day.

I'm such a rule follower by nature (*unless the rule is stupid and, yes, I am the sole arbiter of "stupid"*) that I

---

[212] You should see the way I manage the household budget, hahahaha.

clipped only from plants in the public right-of-way, that space between the street and sidewalk. I pinched off pieces of frilly weeds growing next to the stop signs and intrepid stems pushing through the cracks of the sidewalk (*"these don't belong to anyone!"*).

Then one day, I found myself slipping a pocketknife, the kind with tiny scissors, into my pocket for each foray. Before I knew it, I'd moved on to leaflets peeking out between fence slats and unruly stems poking out from neatly trimmed hedges. *"Their gardener will trim this on their next visit, anyway. I'm saving them money!"*

Even as I clung to my justifications, I realized an onlooker might see it quite differently—as in theft. At first, I looked both ways before clipping, checking for any dog walkers or porch sitters who might rat me out. Till I got caught up in the intrigue, anyway, and I got careless.

And then I got caught. I felt her eyes on me as I turned from the hedge with a fistful of fronds. She sat at the wheel of her car eyeing me in her rearview mirror. Angry homeowner? I froze. Thankfully, she drove away before I passed out from lack of oxygen.

I rushed home without pinching another snippet. We're all so on edge these days, well, I am, anyway, that I imagined the worst possible repercussions. Did her Ring camera film my villainy? Will she post the video on Nextdoor, to alert people to this weird gray-haired plant thief roaming our streets? I guzzled an organic raw kombucha to settle my nerves.

Thing is. I can't quit. Asking me to pass up those tempting lovelies feels as impossible as coming out of an art supply store emptyhanded. Since I'm not aware of any

12-step groups for unrepentant guerrilla foragers, I'm going on the defense. Henceforth, I'll be donning my ginormous sun hat and sunglasses the size of my face before pilfering. I'm also practicing some evasive maneuvers. It's all in the footwork.

I finally screw up the courage to read the email. It's not about me, phew, but I'm heeding this cautionary alert. Time to serpentine.

"It's a beautiful day in the neighborhood, a beautiful day for a neighbor. Would you be mine; could you be mine? Won't you be my neighbor?"
Fred Rogers

TWENTY-EIGHT

Here To Stay: Enduring Turbulent Times

Rivers pour off me as if I'm showering, but my own pores fuel this particular deluge. My lungs strain to wring oxygen from murky thick air. My eyes squint in the blackness to detect the outlines of my many companions. My folded limbs ache, constrained in this cramped enclosure, the roof so low only a toddler could stand upright. The stench excreted by the mass of half-naked bodies, jam-packed in this Hades encased in black plastic, adds a repugnant *je ne sais quoi*. Every survival instinct in my psyche urges me to flee this suffocating space tended by a madman muttering over red-hot rocks.

How could I have ever believed such misery would lead to my spiritual awakening? Only one person crouches between me and the escape hatch. She's small and I think I can take her.

The events that brought me here flash before my burning eyes, like the "life review" of any near-death experience.

"Come to retreat," my teacher had said.

"Go to the desert in the summer?" I'd scoffed. "I can't take the heat."

"You can," she had said.

"But I shouldn't spend the money," I had protested.

"Go," my husband had encouraged.

"But I don't want to go alone," I had pouted.

Then my friend offered to run away with me.

And so, we did.

Standing at the airport departure gate, I'd felt sick to my stomach, racked with guilt for abandoning my small children for this self-indulgence. Until, that is, we'd left the jetways and the interstate behind, cruising the two-lane in our rental car, lured by the massive red rocks on the horizon. Chillaxed at last, we cranked up the Ganesha mantra on the car stereo, chanted 108 times to clear obstacles, and soared into Sedona. A bevy of beauties cheered our arrival at the retreat house, and I fell into welcoming arms, laughing. I vowed to surrender my resistance. *Let's do this.*

Armed with all the water I could carry, I managed to survive the hike in the desert heat that next day by walking at my own stubborn pace and repeating that Ganesha mantra. At nightfall, the drumming of Sara Eaglewoman eviscerated my self-restraint. I writhed and whirled like a dervish on amphetamines.

The following morning, certain that I had no sweat—and no fucks—left to give, I fell in with the group walking to the sweat lodge, as casually as if going to the movies. Upon spotting the low-slung enclosure, encased in black plastic and plastered with heavy blankets, all my resolve evaporated like a bead of sweat on a hot plate. I'd never been claustrophobic before now, but I knew well the symptoms. My mother had suffered mightily from it, too terror-stricken to ride an elevator or an airplane. I felt her panic grab me by the shoulders.

"I'm scared. I can't do this," I whispered to my teacher, then crumpled to the ground before I could faint.

"You're here on Purpose," my teacher soothed.

"I can't, I can't, I can't. . ." I moaned, rocking.

"You can, Cheryl."

"Hmph. Well, I MUST sit by the door," I demanded. I hung back as the brave souls before me crawled into the womb-like space, disappearing like ants soldiering into the compound. I crept in last, trembling and sniffling, but somewhat soothed by the wisp of air floating through the open flap. Then another body thrust through the doorway, and I moved over one space to accommodate the only soul more visibly shaken than I.

It's a consideration I regretted when the fire keeper handed the giant tongs full of glowing red rocks from the sacred fire outside the tent to the elder at the center of the lodge. Nostrils burned, throat singed, and lungs shrieked in protest. When the elder began to drum, my heart pounded in unison, and my freaked-out mind urged me to get out NOW. "This is a huge effing mistake," I whisper-screamed, feeling as I did when my first labor pains began—the horrors to come unknowable but certain to shatter my fragile hold on sanity.

Just when I was sure I'd spend the remainder of my days rocking in a corner, my neighbor, the small one, bolted out the door flap. My heart leapt, as my smoldering mind processed the miracle. But before I could work free my cramped-up limbs, a hand emerged from the tangle of bodies before me and pressed my knee.

"Stay," commanded my teacher from her perch in front of me. In the moment I hesitated, the firekeeper outside the lodge sealed the flap before I could run for it. Groaning, I sidled over, back closer to the exit. Just in case. My hands explored the darkness behind me until I found where the overlying tarp met the ground. I slid my fingers into the divide and raised the covering an inch, maybe two. Not enough to make any discernible difference in the suffocating swelter in the lodge, but enough to give me a sense of agency, access to a finger of air for the absolute worst-case scenario. I began chanting with the tribe and my terror eased. I exhaled all the pent-up paranoia and realized that I was fine.

"I'm going to live."

That first experience in a sweat lodge keeps coming back to me lately. Reading the Daily Deluge of Bad News for Life on Earth inspires the same sort of panic that sweat lodge induced. Because it's not just about the climate crisis. Scientists now say that all of Earth's life support systems are badly damaged. A recent study concluded that six of nine "planetary boundaries" assessed to judge Earth's overall health are completely out of whack.[213] Two of the remaining metrics will surely be breached if we don't act. And still, we humans are still wasting precious time arguing about whether such ominous findings merit any changes to our environment-destroying lifestyles.

---

[213] Sound familiar? I sure hope so. I wrote this essay before the rest of the book.

If my doctor said I was failing in six vital areas essential for life, I'd be in a panic, downing the fruits and veggies, hitting the treadmill, and guzzling the spring water. Yet when Earth's life support systems are failing, we just go on firing up the fossil fuels, mowing down the forests, and flinging microplastics across the globe. As a mother, my heart cannot comprehend such callous irresponsibility. As a lover of my creature comforts, I stand with my hands in the cookie jar while complaining about cookie thieves.

Today, I read in the *New York Times* the grim stories of researchers who are witnessing the extinction of polar bears and penguins, animals they've devoted years to studying. Their concern, candor, and grief left me weeping, heaving, and gasping for air as I was that day in the sweat lodge. It all felt like too much to bear, and I wanted to flee, tune it all out, sink into a vat of wine, or plug my ears and suck my thumb. And then I felt my teacher's hand on my knee. I heard her voice.

"Stay."

Sigh. I unclenched my jaw, chanted a few mantras, hiked a mountain, threw some darts, screamed into my pillow, banged a pizza box, and did whatever I could to slow the descent into paralyzing madness. If only for my children.

Now young adults, my kids stare down their futures at this time of upheaval and dire forecasts with stoic courage. My desperate hope for a planet healthy enough to support their long, full, happy lives urges me to stay present, to be here for and with them. Do what I can, when I can, the best I can, given my time, resources, abilities, and mental health. The many threats to all that we hold dear can either be a nudge to get up off the couch and ACT. Or I can make them my excuse to tune it out, order more crap from Amazon while sipping overpriced craft cocktails. I'd be lying if I said the latter isn't tempting. But it did nothing to improve Earth's health in all the 89 times I tried, so it's time for a more proactive tack. I pray we employ the heat and pressure of these critical times to make diamonds of our carbon-burning excesses.

Turns out, my teacher was right. I can take the heat. It's not easy, and I still keep my seat by the door, my fingers wedged into the divide, ready to release a whiff of cool air to stave off insanity. But when I hold my children's faces in my heart, I know one thing for certain: I'm here to STAY.

THIS IS A PERFECT MOMENT

It's a perfect moment for many reasons, but especially because you and I are waking up from our sleepwalking, thumb-sucking, dumb-clucking collusion with the masters of delusion and destruction.

Thanks to them, from whom the painful blessings flow, we are waking up.

*Their wars and tortures,
their crimes against nature,
extinctions of species
their engineered diseases.*

*Their spying and lying
in the name of the father,
sterilizing seeds and
trademarking water.*

*Molestations of God,
celebrations of shame,
mangling our dreams and
defiling our names.*

*Their ruthless commercials
and blood-sucking hustles,
their endless rehearsals
for the end of the world.*

Thanks to them, from whom the painful blessings flow, we are waking up.

Their painful blessings are cracking open more and more gashes in the sour and shrunken mass hallucination that is mistakenly called "reality." And through the fractures, ripe eternity is flooding in; news of our souls' true home is pouring in; our allies from the other side of the veil are swarming in, inspiring us to become smarter and wilder and kinder and trickier.

We are waking up.

Rob Brezsny
excerpt from *Pronoia Is the Antidote for Paranoia: How the Whole World Is Conspiring to Shower You with Blessings*

TWENTY-NINE
Closing Thoughts: Good Grief

Final playlist: Dreamer's Disease[214]

I find myself on the sidewalk, in front of a weathered red brick shoe-factory-turned-apartment building in downtown Kansas City, Missouri. I lived here a few decades ago, so I am astonished to see someone I recognize walk by. Did they never leave? Or did they wake up here, too?

Before I can decide, I'm transported inside, fumbling through the darkened rooms, gleefully telling my daughter, who is suddenly here, how many stories this place holds. I find my way around the hallways and few rooms with ease, which is so satisfying until I start to wake up and, in the nebulous ethers, I realize I've never lived here. The layout, the décor, the furniture . . . none of

---

[214] Listen on Spotify: https://tinyurl.com/t7y2rnub OR Tidal https://tinyurl.com/4rca6fjv . Or search for "Dreamer's Disease for Madness Readers."

this appears anywhere on my lived timeline. It's both familiar and foreign and my waking self struggles to make sense of it. End scene.

A cup of coffee later, I shrug it off. A close friend is visiting KC next week, that must be the trigger.

Until I sit down and write, then a creeping realization dawns. . . . a world that feels both familiar and foreign sums up my everyday experience these days. Roe v. Wade was settled law for 50 years—until it wasn't. Climate change was something I relied on scientists and the EPA to study and solve. Until ExxonMobil convinced everyone it wasn't worth worrying about. I was certain that basic respect for all life, even when we disagree, was a shared value until social media made it okay to blast death threats for perceived infractions.

Given where we are, not just in the United States, but in the populist and isolationist movements around the world—closing borders, denigrating asylum seekers fleeing torture, scorning environmentalists—I'm doubting humans can work together, as we must, to solve problems on a planetary scale. Doubting that we have the will to drop our tribalism long enough to save ourselves.

Am I back to where I started? Blaming God for the human-caused catastrophes we seem incapable of resolving? It seems a hollow point at this juncture, given the mounting piles of scientific facts and data, the horrific wildfires burning around me, and the heartbreaking pleas of islanders on submerging lands staring into our uncaring eyes.

Just because we have certain innate tendencies doesn't mean we can absolve ourselves of responsibility for the long-term havoc we've caused. We can either lean on our excuses or choose to rise above them with the resolve of Martin Luther King, Jr., Rosa Parks, Wangari Maathai,[215] Malala Yousafzai,[216] and Greta Thunberg.[217] Humans have managed to erect great edifices, build great societies, and solve scientific mysteries over centuries. The Great Wall of China was built in stages over 2,000 years. The Milan Cathedral was erected between 1386 and 1965, some 579 years. Heck, ExxonMobil's disinformation campaign lasted some three uninterrupted decades. We cannot blame our failure to ward off the looming eco-apocalypse solely on the relative "newness" of our prefrontal cortex, which governs our ability to plan for the future.

All of that said, I've had enough of the ranting and despairing. For today, anyway. I wish I could report the writing of this book has eased my fury, cleaned me out, like the prep for a colonoscopy,[218] but I know better. I may look like a peaceful meadow in a pastoral painting—until somebody drops a match on a scorching hot day, and my dry, brown grasses incinerate in a flash. Suffice it to say, I keep a supply of fire extinguishers at my side.

But, at this point, some two-hundred plus pages into this book, we've had enough of all that, haven't we? I

---

[215] Nobel Prize Winner and founder of the Green Belt Movement (see Deforestation chapter).
[216] Advocate for girls' education since she was shot by the Taliban in Pakistan in 2012.
[217] Thunberg's school strikes to demand action on climate change made her an icon of global environmental advocacy. Despite much ridicule and threats, she continues to inspire young activists worldwide to demand action on climate change.
[218] If you have not yet experienced it yet, kids, JUST YOU WAIT.

know I have. I'm so eager for a break from all the mad-shrieking that I actually cleaned the cat box today.

It's that bad.

Music always gives me a welcome break from the insanity in my head, so I fired up my "You Got This!" playlist, and first up is this gem: "You Get What You Give" by the New Radicals. No surprise there, as it's what I routinely shriek-sing while sobbing. The song lifts up those of us who dare to hope, we who have a bad case of the "dreamer's disease."

That's what I've got, I realize with an over-caffeinated jolt: the dreamer's disease. No matter how much I despair, how many times I've cursed the deities for screwing us over, how many times I've pulled the covers over my head . . . I'm here to stay. If only for my children, I'm still in it. I have the dreamer's disease.

So, I created one more playlist for us, entitled "The Dreamer's Disease," to give us all a melodious nudge to stay in the game. Which I celebrated for one self-satisfied moment. Until—and this is where the trouble always comes in—I thought about it.

"*Disease*."

Harkening back to Chapter One, I still flinch to say the word "disease" when describing something I mean to be so positive. Sigh.

But let's explore the word "disease," shall we?

Merriam-Webster says: "When disease was first used, it referred literally to "lack of ease or comfort," as in dis-ease. Well, that sounds Hoover dam apt to me. I am feeling an astronomical "lack of ease" these days, even when I am daring to dream for a better future. I feel a lack of ease just cleaning out the lint tray in the clothes dryer. I've found some truly disturbing stuff in there.[219]

Don't we require some level of dis-ease to imagine a better future? Who even thinks of planning a better way when living in the lap of luxury? "The Dreamers Disease" title stays.

Because as much as "they" want us to believe that it's up to individuals (à la "only YOU can prevent forest fires"[220]), it's simply not true. All of this "it's up to the consumer to do the right thing" bullshit is just one more way of keeping us all isolated and fragmented. Me, scurrying around in my thrift store garb, toting reusable bags, eschewing plastic straws, and fumbling with cardboard tube deodorant, that's the Great Plan for Saving Humanity? That's not even a "concept" of a plan.

Fuck that.

I'm done with the asterisks.

No matter how many reusable straws we use or electric cars we drive, those choices are never going to be enough. They may make us feel better, which is critical for our sense of well-being—and I don't mean to discount the importance of that—but it won't save the coral reefs from bleaching or downsize the superstorms. Fossil fuel companies foisted those lies on us as a sleight of hand, to take our eyes off the truth. "*Look in the mirror, peasants,*

---

[219] And, like Forrest Gump, that's all I have to say about that.
[220] Even as a kid, I called BS on this infamous Smokey the Bear slogan. Hasn't the U.S. Forest Service ever heard of fires caused by lightning strikes?? But it is true that human start 80% of wildfires, and I don't mean to discount our responsibility.

and not at what we're doing to destroy essential life support systems." Grrr. What happens if we just don't buy their lies anymore?

Let me rephrase all that. I have come to believe that, yes, it's important to get our own houses in order, to strap on our oxygen masks first, before rushing out to "save the world." That was the message of my book *Love Earth Now*. And still, I must remind myself daily to practice what I preach, to do that meditation, eat those raw veggies, drink that water, if I hope to have the wherewithal to take out the trash, let alone make a greater contribution. Attention to physical and mental health has never been more vital, when you consider the number of crises we face.

"Hidden within catastrophes are gifts and opportunities; perhaps we have the chance here to awaken potentialities within us, perhaps to live the most meaningful lives that we could ever hope to live."
The Holy Universe by David Christopher

The author of that quote, David Christopher, recently referred me the Good Grief Network, as an outlet for doing some of that oxygen-mask healing. It's a nonprofit that "brings people together to metabolize collective grief, eco-distress, and other heavy emotions that arise in response to daunting planetary crises." I try to ignore the

hunger pangs that arrive whenever I read "metabolize."
To be on the safe side, I grab a handful of cashews, so I'm
not distracted by my growling stomach.

> **GOOD GRIEF NETWORK**
>
> The Good Grief Network facilitates support groups
> designed to help people process heavy emotions,
> transforming fear and fury into meaningful action. Their
> flagship program, "**10 Steps to Resilience &
> Empowerment in a Chaotic Climate**," was inspired
> by the 12-step format of *Adult Children of Alcoholics*
> groups. If you're not sure how to process your own
> furious stew of feelings, perhaps the Good Grief Network
> can provide a framework for your journey.
> http://www.goodgriefnetwork.org/

Once I've corralled my drunk and disorderly ducks
into some semblance of a row, the next step for me is to
link arms with likeminded folks, to remind myself that I
have more to give than just my own paltry efforts.
Countless numbers, SCADS, of people are coming
together in groups of all varieties and sizes and for the
purposes that call to us, challenging the entrenched
systems that keep us swimming in plastics, toxic
chemicals, and fossil fuels. We wield a far greater impact
working in communities, and that's the kind of
supersizing we need if we are to upend the systems that
have normalized "get-mine-before-it's-gone" and

destroying entire ecosystems for hamburgers. We must come together in numbers that matter to manifest a different story for ourselves, to make our voices loud enough to be heard over the drone of a gazillion TVs, tablets, and game consoles.

Brrring brrring, I suddenly hear the '70s calling me, Helen Reddy belting out lyrics about the power of women who roar "in numbers too big to ignore." Yeah. That line keeps coming back to me.

Whether we join a chapter of the Citizens' Climate Lobby, a shareholders' group, or a community garden, we're taking a critical step toward manifesting a more enlightened, compassionate and sustainable human presence on Earth. And we will get there faster working as groups than just swapping out our lightbulbs for LEDs. A 2018 study found that only 25% of a population needs to take a particular stance to flip the majority to the minority position. I pray, for the sakes of our children, and for their children, that it's true because the odds of us convincing everyone of any one thing are as poor as for me giving up list-making.

Still not sure where you fit, what you have to give that will make a difference? Check out the "Changemaker Quiz" on the Story of Stuff website. Being a changemaker doesn't require, much to my relief, spending months in a tree or chained to a tractor. My knees, back, and bladder all say NO.

This quiz identifies the various roles activists play and helps us all to identify our best fit, including Networker, Nurturer, Builder, Investigator, and Resistor. If you're wondering where your role in this stage play of Madness lies, I urge you to check it out. My quiz results identified my role as a Communicator. Surprise surprise! My low back is relieved, knowing I can keep doing my thing supported by an ergonomic pillow.

Identifying my role doesn't mean "problem solved," of course, but it does help me dial down the list I haven't mentioned yet: The Guilt List. It's a hot, steaming pile of all the myriads of things I've failed to do in support of our life support systems. I pick it up when I feel like poking a sharp stick on my eye. Not my happy place.

Focusing on what I can do, what I have to give, on the other hand, takes my gaze off the guilt and onto a happier, somewhat familiar place called HOPE.

"Though it be trite to say, there are no accidents. I am convinced that we are guided to know what to do, if we're willing to ask, to look, listen, and learn. And willing to learn from the most unlikely teachers. Everything we do matters.
All that is born of our love, hope, caring, intuition, faith, imagination, and plain old wishing for something better matters to Earth, to our beloveds, to our communities, to Nature, to our own hearts like glazed donuts matter to Homer Simpson. Even if like Trevor, the young humanitarian in the movie Pay It Forward, we have no clue how or when or to how many, our contributions matter."

Love Earth Now by Cheryl Leutjen

Looking at this excerpt now, I want to make an addition, a single word, one that will coalesce all these thousands of angry and loving words I've written across these two books:

ANGER

As in: "All that is born of our ANGER, love, hope, caring, intuition, faith, imagination, and plain old wishing for something better matters to Earth..." Because it's not just the "nice feelings" that stir us to action. I'm much more likely to get up to rant at neighbors blasting heavy metal music at midnight than to compliment them on their lovely garden.[221]

It's because we LOVE that we get angry in the first place, isn't it? Threaten to harm my children, the ones I love with every fiber in my being, and you'll know the depths of my fury. Threaten to harm that hideous hostess gift I stashed in the back of my closet . . . and I may not even look up from my phone.

Which makes me think there is a place for anger alongside love, when considering our places on this planet at this maddening moment in time. Grabbing a pen and a crumpled, mile-long CVS receipt, I pen:

LOVE EARTH MADLY

Which reminds me of another tune, the Doors wailing about loving her madly. The "her" in the song surely refers to a human, but my mind substitutes another "her." Yes, I do love her, my Earth, like the mad woman I surely am.

---

[221] Sorry about that, neighbors. It truly is lovely.

If you do, too, I urge you to embrace the advice of Shyamala Gopalan Harris, mother of the current U.S. Vice President: [222]

"Don't just sit around and complain. DO SOMETHING!"

Believe you me that I am looking in the mirror when I repeat that refrain. I don't call myself the eco-worrier for nothing. Sure, I tote my gear and put my food scraps out to compost. But my Olympic sport is hand wringing.

The stakes are so high now that I need to up my game. Like a thousandfold. And I mean to gather with my people when I do.

I'm thinking of the playground game of my youth, Red Rover, where we held hands to prevent a runner from the other side from penetrating our line. If we didn't hold tight, we'd lose the game. If Johnny refused to hold hands with Rebecca because, *ew*, cooties? We didn't have a chance. Funny how many of us (supposed) adults, me included, are having those "ew, cooties" moments when thinking about people who've been assigned a different team color, red v. blue.

It's time to do what humanity has done throughout the ages when faced with insurmountable odds, like the early humans staring down the megafauna animals of the Pleistocene Epoch[223]—the mastodons, the killer saber-

---

[222] Again, I was praying that Kamala Harris's title by the time you read this would be Madame President, but instead, I've been given a lot more to complain about, I *mean*, DO.

[223] The Pleistocene Epoch lasted from about 2.5 million to 11,700 years ago. Early humans evolved about 2 million years ago, and they were smaller than us, so just imagine waking up to those giants every day.

toothed cats, and the glyptodon, an armadillo-looking critter the size of a VW Bug. Yes, an armadillo the size of a small CAR.

We clasp hands, lock arms, gather around the sacred fire, share our stories, say our prayers or mantras, and make our plan of attack. Then we, gather the tools of our era, and go at that killer saber-toothed cat with all we've got.

"We are in a tough spot. We find ourselves desperately looking to politicians worldwide to lead us through this catastrophe. Yet vested interests from multinational corporations maintain a stranglehold on top-down change... This is where community-level action is a catalyst for creating equitable, life-centered ways of being."

Excerpt from *How to Live in a Chaotic Climate: 10 Steps to Reconnect with Ourselves, Our Communities and Our Planet*
by LaUra Schmidt, with Aimee Lewis Reau and Chelsie Rivera

THIRTY

Going Forth

It's Game Day, folks.

I'm no expert in the rules and strategies of the sporty games, but I've cheered with the husband for enough Kansas City Chiefs football games to glean this:

Star Quarterback Patrick Mahomes is in a lot of commercials. It's a shame he isn't paid enough to make it on his football salary, but I sympathize. I've had to pick up a second or third gig at times, too.

Wait, that wasn't my actual point. What I've also gathered from my many hours of being in the same room as someone watching football is this: all the teams have something called a game plan. From what I can surmise, the game plan consists of a bunch of plays that the teams practice over and over. That way, when the next bad thing happens, they have a plan, a play to *execute*, instead of just crying on the floor of their closets. Smart.

There are trick plays, too, which our buddy Mahomes seems to favor. My favorite was when the offense circled up and did a ring-around-the-rosy. Those big hulking dudes are more graceful than you'd think. Then, just when you think they'd be too dizzy, somebody throws the ball to someone and it's back to football as I least understand it!

American football—and that's something I've learned to say since watching *Ted Lasso*—is a grueling sport. Players risk blown-out knees, traumatic brain injuries and getting traded to Buffalo in the winter. Nothing against Buffalo, but there's no dome over that stadium, brrrrrr.

It's good to see they can have some fun out there, too, even when the chips are down. Way down. Because get this: there are even plays for doomsday scenarios like the Chiefs being so far down, they had zero points and Houston had 42. Mahomes tells the team to buck up and run the plays in their game plan, and they end up winning the game.

What determination! If I'd been there, my team down 42 points, I'd be packing up. "Grab your coats, kids," and get an early start out of the stadium to beat the traffic.

But that's not really an option now, is it? Not if we want to leave any sort of inhabitable planet behind for our children, nieces and nephews, grandchildren, godchildren, and all the young people who showed up on the playing field with just seconds remaining on the game clock, already down by a bazillion points, and the referees ruling against them.

Time to cue up the calm-under-pressure of Patrick Mahomes, huddle up with our teams, and run whatever Hail Mary plays we've got in our playbooks. We need every player on that field giving it all they've got.

It's Game Day, folks. The Big Game day, the one that decides the outcome of this entire season of humanity.

We came here to play.
Full out.
With *everything* on the line.

Player One, ready?

"Hard times require furious dancing."
Alice Walker

Book Club Questions

1. Was there a particular story or quote that particularly resonated with you? Why do you think it impacted you?

2. Did you find any part of *Madness on the Brink of Eco-Apocalypse* unexpected or surprising? What took you by surprise and why?

3. The author shares her way of coping with "eco-angst." How did this influence your thoughts on managing your mental health, relationships, or productivity when dealing with distressing news about the future of our planet?

4. Reflect on the author's decision to invoke her "personal faith crisis." How does Leutjen's perspective on the intersection of personal well-being and environmental concern challenge or align with your own experiences?

5. Discuss the implications of human decision-making that often contradicts our long-term self interest. Leutjen states, "Humans take pride in our intelligence as the thing that sets us above

and apart from other species, and yet I don't see the other species lighting their homes on fire and calling it God's will or essential for 'progress.'" How do you think political divisions, especially in the United States, have affected our ability to enact laws to protect the planet?

6. After reading this book, are you inspired to take action to reduce your environmental impact or join a community of people making a difference? What actionable steps will you identify for yourself or your community?

Acknowledgments

I'm thankful for the many people who tolerated, *I mean*, supported me in the process of writing this maddening book. When days of research into the dark truths plunged me into the depths of despair and hopelessness, I drew from your strength, your wit, and your wine to forge on. Please send thoughts and prayers for my husband, David, and son, Cameron, who had to live with me through it all.

I'm especially grateful to advance readers Reni Fulton, Davina Kotulski, Nina Lesowitz, Daniel Lott, Kathy McCourt, Cynthia Townsend, Tracy Warriner, and Jennifer Weissmuller.

I send SCADS of gratitude to story editor and BS detector Christy Kickass Bell. This "book" was a chaotic mess, much like my state of mind when writing it, before she got her capable hands on it.

I'm grateful for Nita Sweeney's generosity of time and expertise on all matters relating to book publishing. Her wisdom as a mental health advocate, combined with her knowledge as the author of four books, helped me get this infuriating project over the finish line. Get your own dose of mindfulness from one of her books: https://nitasweeney.com/

I'm beyond words thankful for my book cover

designer and daughter, Chloë Meyer, especially for her patience with my myriad changes, including a very last-minute title change. If you ever need a graphic designer who takes the time to understand your needs, I highly recommend her work—and I'm not just saying that because I'm her mom.

www.chloe-meyer.com

I extend my heartfelt gratitude to all the scientists who conduct the essential yet often heartbreaking research that reveals the harmful effects of our environment-destroying habits. How they muster the fortitude to persist in the face of such dire truths is beyond my comprehension. They deserve on-our-knees reverence and accolades instead of threats and insults.

Above all, I thank the people featured in this book for taking the time to contribute quotes and to share their work. Whenever I am convinced "nobody cares," know that I re-read your words, your commitments to caring for our Earth and all her of EarthKind, and my reservoir of hope for humanity refills to overflowing. I thank you all for showing up and Doing the Work of caring for this beloved Earth home of ours.

Note From The Author

This book almost didn't happen. The title and the rancor seemed a stark reversal of everything I said in *Love Earth Now*. In that book, I wrote about finding peace in a world gone mad. Would it be contradictory to rant and rave in anger about the same mad world? Would I risk bad karma or lightning bolts from on high if I expressed my frustration with our Creator?

I enjoy a deep and devoted spiritual practice, and, much as I disparage our circumstances, my devotion to our Creator is unwavering. That said, I have my moments of doubt and frustration, as does anyone with a thoughtful relationship with a higher power, I suppose.

I do thank you for taking the time to read this book. If you feel that others would benefit from anything you found meaningful on these pages, I welcome your positive feedback on social media, on the Amazon listing, on Goodreads or in personal emails. We need each other now more than ever. Tell me how you're navigating these maddening times.

Facebook: @LoveEarthNow
Instagram: @LoveEarthNow
Etsy shop: https://loveearthnow.etsy.com

Email: Cheryl@LoveEarthNow.us
Newsletter signup: http://eepurl.com/dn5e-z
Websites: https://cherylleutjen.com/,
https://loveearthnow.us/

About The Author

Cheryl Leutjen draws from her experience as a geologist, attorney, small business owner, spiritual practitioner, nervous stand-up comic, and worried mother. This wealth of experience, along with degrees in Interdisciplinary Ecology, Environmental Geology, and Law, affords her a broad perspective to contemplate the dire environmental challenges of our time. She writes, rants, and facilitates creative experiences to inspire us all to discover our own path of living more Earth-mindfully.

Her book, *Love Earth Now*, won a Silver Nautilus Book Award.

She lives in Los Angeles with her husband and three cats who care not one whit about her credentials.

Also By Cheryl Leutjen

Love Earth Now:
The Power of Doing One Thing Every Day

Do you find yourself wondering what on earth you can do about the scary-sounding environmental challenges of our time? Do you wish you could do something to make a difference, but doubt you have the time, energy, money, or power?

Love Earth Now is your go-to guide for discovering what you can do to effect meaningful change—starting right now. Author Cheryl Leutjen's wise book of planetary self-help is a deeply thoughtful and often comedic exploration of her own efforts to make an eco-contribution. Through personal observation that she records in stunningly beautiful prose, Leutjen's ode to our planet is one of the most distinctive ecological books to come along in a generation.

Each chapter of *Love Earth Now* concludes with a planet-positive "Love Earth Invitation," a simple and immediate exercise that prompts you to explore your own feelings and calls to action. These eco-mindfulness moments provide the opportunity to reflect and discover what you can do right now to contribute to a sustainable future for us all.

See LoveEarthNow.us for more information.

Praise For Love Earth Now

"In her debut book, author Cheryl Leutjen, brings to life the internal voice that so many of us find ricocheting in the universe of minds, our very beings. She lovingly slows us down so that we might be forgiving, on so many levels: with ourselves, with the state of affairs, with our humanity—which is forever a journey of the (s)hero."

- **Reverend Dr. Kate Rodger, Founder of the Institute of Modern Wisdom and Agape-Ordained Minister**

"This is not your typical '50 things you can do" book'—it's a memoir of what it's like to be awake and open to the wrenching reality of living in a society seemingly bent on destruction (including self-destruction), to feel the pain one feels hearing story after story of what we're doing to the Web of Life, and to face the difficulty of finding the nearly impossible: living sustainably in this unsustainable society.

She offers invitations and meditations that can help you clarify your path, but I found her greatest gift toward the end of the book—a wonderfully unexpected story about giving up, where you reach the point where you can't go on, you can't take in any more . . . you can't just do it anymore."

-David Christopher, author of *The Holy Universe* (Nautilus Book Awards Silver winner, USA Best Books Award winner) and creator of *The Smallest Vessel* Project.

References & Resources

INTRODUCTION

"**climate scientists surveyed**." https://www.globalwitness.org/en/campaigns/digital-threats/global-hating/
"**original meaning of 'woke'**." Montanaro, Domenico, "Republicans can't stop using the word 'woke'. But what does it really mean?" NPR.org. July 21, 2023. https://www.wwno.org/npr-news/npr-news/2023-07-21/republicans-cant-stop-using-the-word-woke-but-what-does-it-really-mean
"**drinking bleach**." Qamar, Aysha. "At least 5 states report an increase in calls to poison control after Trump's 'disinfectant' COVID-19 remarks." Daily Kos, February 2020. April 28, 2020. https://www.dailykos.com/stories/2020/4/28/1941065/-States-report-an-increase-in-calls-to-poison-control-after-Trump-s-disinfectant-COVID-19-remarks Accessed May 12, 2024
"**2023 hottest year**." NOAA, "2023 was the world's warmest year on record, by far." January 12, 2024. https://www.noaa.gov/news/2023-was-worlds-warmest-year-on-record-by-far Accessed May 9, 2024.
"**Secretary of Defense quote**." https://www.defense.gov/News/News-Stories/Article/Article/2582051/defense-secretary-calls-climate-change-an-existential-threat/

CHAPTER ONE: MAD AT GOD

"**President Bush known for preferring stability**." See "Remembering George H.W. Bush, the "Environmental President" By Amaury Laporte, December 5, 2018. https://www.eesi.org/articles/view/remembering-george-h.w.-bush-the-environmental-president .
"**President Bush quote**." Comments at the 1992 Earth Summit in Rio de Janciro. Laporte, Remembering George H.W. Bush. https://bush41library.tamu.edu/archives/public-papers/4417
"**U.S. Global Change Research Program**." https://www.globalchange.gov/about-us
"**Fossil fuel subsidies**." https://www.imf.org/en/Blogs/Articles/2023/08/24/fossil-fuel-

subsidies-surged-to-record-7-trillion
"**Drilled Podcast**." https://drilled.media/podcasts/drilled
"**Our Biggest Fear**." Helen Ubiñas, "Our Biggest Fear is Each Other." Philadelphia Inquirer, April 19, 2023. https://t.co/afQpNMcaFH Accessed May 9, 2024.
"**Rex Tillerson quote**." ExxonMobil shareholder meeting, May 29, 2013. Joe Romm, "Exxon CEO: What Good Is It to Save the Planet If Humanity Suffers?" *ThinkProgress*, May 30, 2013, https://archive.thinkprogress.org/exxon-ceo-what-good-is-it-to-save-the-planet-if-humanity-suffers
"**get-out-of-the-way machines**." Daniel Gilbert. "Global Warming and Psychology." Harvard Thinks Big 2010. Vimeo video, 18:32. Posted by Harvard University, February 22, 2010.
https://vimeo.com/10324258
"**Pacific Islander emissions**."
https://www.weforum.org/agenda/2024/05/small-island-states-making-big-strides-towards-net-zero/#
"**Paycheck Protection Loans forgiven**." Richards, Zoe. "White House shines light on Republicans who are criticizing student debt cancellation after getting their PPP loans forgiven." NBC News, Aug. 25, 2022.
"**Saleemul Huq quote**." Emmanuel Cappellin, dir. *Once You Know*. France: Pulp Films, 2020.
"**Global warming is deadly threat**" quote. Daniel Gilbert. 'Global Warming and Psychology' at Harvard Thinks Big 2010.
https://vimeo.com/10324258
"**we've got no free will**." Purtill, Corinne, "Stanford scientist, after decades of study, concludes: We don't have free will." Los Angeles Times, 17 October 2023.
https://www.latimes.com/science/story/2023-10-17/stanford-scientist-robert-sapolskys-decades-of-study-led-him-to-conclude-we-dont-have-free-will-determined-book Accessed November 1, 2023.
"**The Divine**" **quote**. Wyckoff, Mallory. God Is. Grand Rapids, MI: Wm. B. Eerdmans Publishing, 2022.
"**Citizens' Climate Lobby**." https://citizensclimatelobby.org/

CHAPTER TWO: TIPSY TIGHTROPE

"**Perkins quote**." Marsh, George Perkins. Man and Nature: Or, Physical Geography as Modified by Human Action. New York: Charles Scribner, 1864.
"**heat wave records**." https://theconversation.com/extreme-heat-is-breaking-global-records-why-this-isnt-just-summer-and-what-climate-change-has-to-do-with-it-234249
"**average heat wave season**."
https://www.smithsonianmag.com/science-nature/six-big-ways-climate-change-could-impact-the-united-states-by-2100-180983547/
"**heat wave 49 days longer.**" Gramling, Carolyn. "Climate Heat: 2023 Set to Be the Hottest Year on Record." Science News. October 2, 2023. https://www.sciencenews.org/article/climate-heat-hottest-year-

record-2023?form=MG0AV3.
"**Peter Kalmus quote**." @ClimateHuman on platform X, Dec 15, 2023.
"**Bureau of Linguistical Reality**."
https://bureauoflinguisticalreality.com/

CHAPTER THREE: BAND PLAYED ON

"**red wolf named Muppet**." Center for Biological Diversity, "Beloved Red Wolf Killed by Vehicle Strike in North Carolina," news release, May 2, 2024, https://biologicaldiversity.org/w/news/press-releases/beloved-red-wolf-killed-by-vehicle-strike-in-north-carolina-2024-05-02/
"**Earth's operating systems failing**." Borenstein, Seth. "Earth is outside its 'safe operating space for humanity' on most key measurements, study says." AP News, Sept 13, 2023. https://apnews.com/article/earth-climate-change-biodiversity-environment-pollution-c8582c3ae0344b5a88cc38cd8e725702 Accessed May 12, 2024.
"**Johan Rockstrom quote**." Ibid.
"**Members of the Titanic band**." Turner, Steve. *The Band that Played On*. Nashville, TN: Thomas Nelson, Inc., 20131.
"**songs played**." Turner, *The Band Played On*, 7-13.
"**Transition Town quote**." https://transitionnetwork.org
"**Lynn Hartle TED Talk**." Hartle, Lynn. "Transition Towns." Filmed November 2021 at TEDxPSUBrandywine. TED video, 19:30. January 2022.
https://www.ted.com/talks/lynn_hartle_transition_towns_jan_2022

CHAPTER FOUR: BIODIVERSITY

"**Recommended Musical Pairing**." Gates-Bahlhorn, Christian James. "What's It All Mean?" Songtrust Avenue, 2023.
"**Earth Out of Bounds Report**." Richardson, Katherine et al. "Earth beyond six of nine planetary boundaries." Science Advances, 13 Sep 2023. https://www.science.org/doi/10.1126/sciadv.adh24583
"**biodiversity defined**."
https://education.nationalgeographic.org/resource/biodiversity/
"**one million species**." Gramlin, Carolyn. "1 million Species are Under Threat." Science News, MAY 8, 2019.
https://www.sciencenews.org/article/1 million-species-under-threat-humans-speed-extinction
"**83% of all wild mammals**." Carrington, Damian. "Humans just 0.01% of all life but have destroyed 83% of wild mammals – study," The Guardian, 21 May 2018.
https://www.theguardian.com/environment/2018/may/21/human-race-just-001-of-all-life-but-has-destroyed-over-80-of-wild-mammals-study
"**cataclysmic episodes**." Ritichie, Hanna. "There have been five mass

extinctions in Earth's history." Our World in Data, November 20, 2022. https://ourworldindata.org/mass-extinctions
"**speeding up the usual rate of extinctions**." Gramlin, Carolyn. "1 Million Species Under Threat. Here are 5 Ways We Speed Up Extinctions." Science News, May 8, 2019. https://www.sciencenews.org/article/1-million-species-under-threat-humans-speed-extinction Accessed May 23, 2024.
"**royal inbreeding**." https://historycollection.com/16-royals-who-suffered-from-hereditary-mutations-and-defects-caused-by-inbreeding/
"**Fran Pavley footnote**." I urge you to read about Fran Pavley's many contributions to progressive environmental policies. https://calmatters.org/environment/2015/07/sen-fran-pavley-a-quiet-force-in-the-climate-storm/
"**cougar deaths**. "Kat Kerlin. "Mountain Lion Mortality Maps Show Rough Road for Cougars." UC Davis News, February 02, 2023. https://www.vetmed.ucdavis.edu/news/mountain-lion-mortality-maps-show-rough-road-cougars
"**habitat too small**." https://web.achive.org/web/20220424102606/https://www.nps.gov/articles/000/puma-profiles-p-022.htm
P-22's legacy." Learn more about P-22 and his legacy at: The Mountain Lion Foundation, https://mountainlion.org/2023/12/15/p22-legacy/
"**cost of wildlife collisions**." https://www.thecgo.org/research/wildlife-crossing-ahead-costs-and-benefits-of-avoided-collisions/
Wallis Annenberg quote." https://annenberg.org/initiatives/wallis-annenberg-wildlife-crossing/

CHAPTER FIVE: MAKE IT HARD

"**Schmidt, LaUra et al quote**." LaUra Schmidt, with Aimee Lewis Reau and Chelsie Rivera, *How to Live in a Chaotic Climate: 10 Steps to Reconnect with Ourselves, Our Communities and Our Planet* (Shambhala Publications, 2023).
"**Paul Hawken quote**." Hawken, Paul. "The Commencement Address to the Class of 2009." University of Portland, May 3, 2009. https://www.commondreams.org/views/2009/05/23/paul-hawkens-commencement-address-class-2009 .

CHAPTER SIX: PLASTICS

"**Recommended Musical Pairing**." Stevens, Cat. "Where Do the Children Play?" Track 1 on *Tea for the Tillerman*. Island Records, 1970. Vinyl.
"**Babies v. Plastics quote**." Charron, Aiden, and Jacob Wunsh. "Babies Vs. Plastics." Earthday.Org. November 21, 2023. https://doi.org/https://www.earthday.org/babies-vs-plastics-what-

every-parent-should-know/
"**The Anthropocene book by John Green**." Green, John. *The Anthropocene Reviewed: Essays on a Human-Centered Planet*. New York: Dutton, 2021.
"**Year of No Garbage book**." Schaub, Eve O. *A Year of No Garbage*. New York: Skyhorse Publishing, 2023.
"**The Fraud of Plastic Recycling report**." Climate Integrity. "Plastics Fraud." Climate Integrity. Accessed September 26, 2024. https://climateintegrity.org/projects/plastics-fraud .
"**The Plastics Pipeline report**." Lerner, Sharon. "The Plastics Pipeline: A Surge of New Production Is on the Way." Yale Environment 360. February 19, 2020. https://e360.yale.edu/features/the-plastics-pipeline-a-surge-of-new-production-is-on-the-way
"**16,000 chemicals**." PlastChem Project. "Home." *PlastChem Project*. Accessed September 26, 2024. https://plastchem-project.org/.
"**chemicals in water in plastic bottles**." **Hornbek, Maria.** "Soft Plastic Bottles Put Over 400 Chemicals into Water." *Futurity*. Last modified February 11, 2022. https://www.futurity.org/soft-plastic-water-bottles-2696062-2/ .
"**nanoplastics in breast milk**." Babies vs. Plastics Report. EarthDay.org, NOVEMBER 21, 2023. https://www.earthday.org/babies-vs-plastics-what-every-parent-should-know/ Accessed May 13, 2024.
"**clogged arteries**." Watson, Clare. "Microplastics Found in Blood Clots in Heart, Brain, And Legs." ScienceAlert.Com. May 23, 2024. https://doi.org/https://www.sciencealert.com/microplastics-found-in-blood-clots-in-heart-brain-and-legs
"**Brittany Watts case**." https://www.nbcnews.com/news/us-news/black-woman-ohio-was-charged-miscarrying-bathroom-experts-warn-dangero-rcna130649
"**recycled plastic toys**." IPEN. "Toys Made with Recycled Plastics Expose Children to High Levels of Toxic Chemicals." *Environment International*, November 16, 2023. https://ipen.org/news/toys-made-recycled-plastics-expose-children-high-levels-toxic-chemicals.
"**firearms deaths**." Matt McGough, Krutika Amin, Nirmita Panchal, and Cynthia Cox. "Child and Teen Firearm Mortality in the U.S. and Peer Countries." Jul 18, 2023. https://www.kff.org/global-health-policy/issue-brief/child-and-teen-firearm-mortality-in-the-u-s-and-peer-countries/
"**fraud of plastics recycling**." Center for Climate Integrity. "The Fraud of Plastic Recycling: How Big Oil and the plastics industry deceived the public for decades and caused the plastic waste crisis." February 2024. https://climateintegrity.org/uploads/media/Fraud-of-Plastic-Recycling-2024.pdf
"**Plastics Pipeline report**." Gardiner, Beth. "The Plastics Pipeline: A Surge of New Production Is on the Way." *Yale Environment 360*, December 19, 2019. https://e360.yale.edu/features/the-plastics-pipeline-a-surge-of-new-production-is-on-the-way .
"**27 more coal plants**." Beyond Plastics. "The New Coal." Beyond Plastics. Accessed September 26, 2024. https://www.beyondplastics.org/publications/the-new-coal.

"**Airplane movie**." Zucker, Jim, David Zucker, and Jerry Zucker, directors. *Airplane!*. Paramount Pictures, 1980.
"**California sues ExxonMobil**." Denise Chow, "California Sues ExxonMobil for Plastic Recycling Deception," NBC News, September 22, 2024, https://www.nbcnews.com/science/environment/california-sues-exxonmobil-plastic-recycling-deception-rcna172267 .
"**Ridwell**." https://www.ridwell.com/transparency
"**Hydroblox**." https://www.hydroblox.com/
"**Movie WALL-E**." Stanton, Andrew, director. *WALL-E*. Walt Disney Home Entertainment, 2008.
"**Plastics Pollution Coalition**."
https://www.plasticpollutioncoalition.org/about
"plastic toys contain hazardous waste." Clemence Budin et al., "Detection of High PBDD/Fs Levels and Dioxin-Like Activity in Toys Using a Combination of GC-HRMS, Rat-Based and Human-Based DR CALUX Reporter Gene Assays," Chemosphere 251, July 2020, https://www.sciencedirect.com/science/article/pii/S0045653520307724
"**Beyond Plastics local group**." https://www.beyondplastics.org/act
""**Cheese Packaged in Plastic**." Beyond Plastics. "Cheese Packaged in Plastic May Expose You to Harmful Chemicals." Accessed September 7, 2024. https://www.beyondplastics.org/fact-sheets/plastic-wrapped-cheese
"**Plastic Wars**," Frontline (2020)
Altman, Rebecca. "A Return to Plastics: A Personal Journey Through Time." 2019.
Heinrich Boell Foundation. The Plastic Atlas. 2021
Schaub, Eve O. *Year of No Garbage: Recycling Lies, Plastic Problems, and One Woman's Trashy Journey to Zero Waste* (p. 313). Skyhorse. Kindle Edition.

CHAPTER SEVEN: BAMBOO FORK

CHAPTER EIGHT: DEAD ZONES

"**Recommended Musical Pairing**." Tom Paxton. "Whose Garden Was This?" Track 1 on *Whose Garden Was This*. Elektra Records, 1970. Vinyl.
"**Gulf of Mexico dead zone**." "Gulf of Mexico 'Dead Zone' Largest Ever Measured." Aug 8, 2017.
https://coastalscience.noaa.gov/news/gulf-mexico-dead-zone-largest-ever-measured/
"Dean Blanchard quote." "The Danger Downstream." Grist.org. Jan 28, 2020. https://grist.org/food/gulf-shrimpers-fight-for-their-livelihoods-in-a-fertilizer-fueled-dead-zone/
"**Dean Blanchard business**."
https://americanshrimp.com/suppliers/dean-blanchard-seafood/
"**costs of dead zones**." Chechinger, Anne. "The High Cost of Algae Blooms in U.S. Waters." Ewg.Org. August 26, 2020.

https://doi.org/https://www.ewg.org/research/high-cost-of-algae-blooms.
"**Find your CSA**." https://www.localharvest.org/csa/
"**Future Farmers**." https://www.ffa.org/about/
"**4-H**." https://4-h.org/

CHAPTER NINE: COOKING UP THERAPY

"**child labor**." https://www.dol.gov/agencies/ilab/our-work/child-forced-labor-trafficking/child-labor-cocoa
"**Food miles**." https://foodwise.org/learn/how-far-does-your-food-travel-to-get-to-your-plate/
"**craftivism**." Corbett, Sarah. *How to Be a Craftivist: The Art of Gentle Protest*. London: Unbound, 2017.
"**Society of Fearless Grandmothers of Santa Barbara**."
https://www.fearlessgrandmotherssb.org/

CHAPTER TEN: DEFORESTATION

"**Recommended Musical Pairing**." Hawley, Richard. "Heart of Oak." *Further*, BMG, 2019.
"**Wall Drug**." https://www.walldrug.com/
"**Forest biodiversity**."
https://openknowledge.fao.org/server/api/core/bitstreams/dfb12960-44ee-4ddc-95f7-bec93fbb141e/content
"**Devouring the Rainforest quote**." "Devouring the Rainforest." *Washington Post*, April 29, 2022.
https://www.washingtonpost.com/world/interactive/2022/amazon-beef-deforestation-brazil/
"**Black Friday explained**." https://www.britannica.com/story/why-is-it-called-black-friday
"**Agriculture and forest clearing**."
https://openknowledge.fao.org/server/api/core/bitstreams/dfb12960-44ee-4ddc-95f7-bec93fbb141e/content
"**European Union Deforestation Regulations**."
https://environment.ec.europa.eu/topics/forests/deforestation/regulation-deforestation-free-products_en
"**Amazon greenwashing**."
https://changingmarkets.org/report/feeding-us-greenwash-an-analysis-of-misleading-claims-in-the-food-sector/
"**The Bonn Challenge**." https://www.bonnchallenge.org/about
"**Field of Dreams movie**." Robinson, Phil Alden, director. *Field of Dreams*. Universal Pictures, 1989.
"**Sarah Lake quote**." Lake, Sarah. "The Hidden Forces Behind Your Food Choices." Filmed June 2024. TED video, 8:30.
https://www.ted.com/talks/sarah_lake_the_hidden_forces_behind_your_food_choices/transcript?subtitle=en
"**Grass-fed beef nutrition**." https://www.webmd.com/diet/grass-fed-beef-good-for-you

"**Stanford study on beef substitution**." Heller, Martin C., et al. "Simple dietary substitutions can reduce carbon footprints and improve dietary quality across diverse segments of the US population." Nature Food 4, no. 9 (2023): 864-873. https://doi.org/10.1038/s43016-023-00864-0.
"**Teton Waters Ranch**." https://tetonwatersranch.com/
"**Creekstone Farms**." https://creekstonefarms.com/
"**Cream Co. Meats**." https://creamcomeats.com/
"**Stemple Creek Farms**." https://stemplecreek.com/
"**Keller Meats**." https://kellermeats.com/
"**North East Trees of Los Angeles**."
https://www.northeasttrees.org/
"**Green Belt Movement**."
https://www.greenbeltmovement.org/who-we-are
Barack Obama quote."
https://www.greenbeltmovement.org/node/307
"**Forest Guide Jackie Kuang**." http://jackiekuang.com/

CHAPTER ELEVEN: THE HATE LIST

"*Ice Age and taxes*." Comments made by U.S. Representative Marjorie Taylor Greene before Congress on April 26, 2023. @Acyn, X.com, April 26, 2023, 7:19pm.
https://twitter.com/Acyn/status/1651410740618862592
"**China's new coal-fired plants**." "China's coal power spree could see over 300 coal plants added before emissions peak," Global Energy Monitor, August 28, 2023.
"**Abraham-Hicks.**"
https://youtu.be/O9zbH5MZSHA?si=I8Ar7OxlREHQMtjD
"**alternative facts**." https://www.nbcnews.com/meet-the-press/meet-press-01-22-17-n710491
"*1811 dictionary*." Grose, Francis. *1811 Dictionary of the Vulgar Tongue*. London: Printed for C. Chappel, 1811.
"**Kevin Monahan**." Collins, Dave. "Man Who Shot Kaylin Gillis Grew Increasingly Bitter About Trespassers, Neighbor Says." *AP News*, April 18, 2023. https://apnews.com/article/fatal-shooting-wrong-driveway-new-york-36fd62216cf3d9c20333d760148269a0.
"**Drilled podcast**." Drilled is a true-crime podcast, hosted by investigative journalist Amy Westervelt who investigates the various obstacles that have prevented the world from adequately responding to climate change. https://drilled.media/podcasts/drilled
"**Dark money news**." Sara Fischer. "Dark money news outlets outpacing local daily newspapers." Axios.com. Jun 11, 2024.
https://www.axios.com/2024/06/11/partisan-news-websites-dark-money
"**CEO salaries**." Bivens, Josh, and Jori Kandra. "CEO Pay in 2022." *Economic Policy Institute*, September 21, 2023.
https://www.epi.org/publication/ceo-pay-in-2022/
"**@Merman_Melville Tweet**."
https://x.com/Merman_Melville/status/1364000670760669184

CHAPTER TWELVE: FRESHWATER

"**Recommended Musical Pairing**." Young, Neil. "Be the Rain." *Greendale*. Reprise Records, 2003. Vinyl.
UNESCO. UN World Water Development Report 2024: Water for Prosperity and Peace. Paris: UNESCO, 2024.
"**cost of tap water**." "How Much does US Tap Water Cost?" America Explained. Last modified May 17, 2024.
https://www.americaexplained.org/how-much-does-us-tap-water-cost.htm
"**Barbara Kingsolver quote**." Kingsolver, Barbara, Steven L. Hopp, and Camille Kingsolver. Animal, Vegetable, Miracle: A Year of Food Life. New York: HarperPerennial, 2008.
"**average price of oil.**" "Crude Oil Price History Chart." Macrotrends. Accessed September 8, 2024.
https://www.macrotrends.net/1369/crude-oil-price-history-chart
"**can't eat money**." Often attributed to the Cree People, this saying was first published in 1972 in a quote by Alanis Obomsawin, Abenanaki from the Odanak reserve, in "Who is the Chairman of this Meeting."
https://quoteinvestigator.com/2011/10/20/last-tree-cut/
"**water use in fracking**." Tabuchi, Hiroko. "Fracking's Hidden Toll on U.S. Water Supplies." The New York Times. September 25, 2023.
https://www.nytimes.com/interactive/2023/09/25/climate/fracking-oil-gas-wells-water.html
"**water use by agriculture**." UNESCO. UN World Water Development Report 2024: Water for Prosperity and Peace. Paris: UNESCO, 2024. https://www.unwater.org/publications/un-world-water-development-report-2024
"**regions of high water stress**." Ceres. "Hydraulic Fracturing & Water Stress: Water Demand by the Numbers." Ceres, accessed September 9, 2024.
https://www.ceres.org/resources/reports/hydraulic-fracturing-water-stress-water-demand-numbers
"**Wintergarden, Texas**." Tabuchi, Hiroko. "Fracking's Hidden Toll on U.S. Water Supplies." The New York Times. September 25, 2023.
https://www.nytimes.com/interactive/2023/09/25/climate/fracking-oil-gas-wells-water.html
"**Cadillac Desert quote**." Reisner, Marc. Cadillac Desert: The American West and Its Disappearing Water. Penguin Publishing Group. 1993.

CHAPTER THIRTEEN: MY UNBUCKET LIST

"**Perfect Days movie**." Wenders, Wim, director. *Perfect Days*. Tokyo: Master Mind Ltd., 2023. Film.
"**Jack Keel quote**." Kheel, Jake. *Waking the Sleeping Giant: Unlocking the Hidden Power of Business to Save the Planet*. Austin, TX: Lioncrest Publishing, 2021.
"**Mary Shaughnessy story**." Shaughnessy, Mary. "A Love Letter to My People." "The Moth" Podcast audio, recorded Nov. 30, 2023,

original air date Sep 10, 2024. https://themoth.org/stories/a-love-letter-to-my-people
"**Green Acres Chocolate Farm and Nature Preserve in Bocas del Toro, Panama**." Planet Rehab. https://www.planetrehab.org/
President Obama quote." Keith, Tamara. "*A Less-Restrained Obama Finally Says 'Bucket'*." NPR.Org. June 29, 2015. https://doi.org/https://www.npr.org/sections/itsallpolitics/2015/06/28/418152881/a-less-restrained-obama-finally-says-bucket
"**Sustainable travel**." Portnoy, Susan. "Travel Responsibly: A Guide for Sustainable Hotels and Tours." *The Insatiable Traveler*. Accessed September 10, 2024. https://theinsatiabletraveler.com/travel-responsibly-guide-for-sustainable-hotels-tours/

CHAPTER FOURTEEN: CLIMATE CHANGE

"**Recommended Musical Pairing**." Webby, Chris. "Our Planet." Featuring Bria Lee. *Wednesday After Next*. EightyHD, 2019.
"**Climate Change Timeline**." BBC. "A brief history of climate change - BBC News." A brief history of climate change. Accessed June 18, 2024. https://www.bbc.com/news/science-environment-15874560.
https://www.bbc.com/news/science-environment-15874560
"**growing certain crops**." https://today.duke.edu/2022/04/how-climate-change-changing-latin-america
"**17 million in Latin America**." https://today.duke.edu/2022/04/how-climate-change-changing-latin-america
"**Gabriela Nagle Alverio quote**." "Student Gabriela Nagle Alverio Describes Her PhD Research on Climate-Induced Migration and Health." Sanford School of Public Policy, Duke University. Accessed September 25, 2024. https://sanford.duke.edu/story/student-gabriela-nagle-alverio-describes-her-phd-research-climate-induced-migration-and/
"**Lake Michigan amount of melt**." "Antarctica and Greenland Are Melting Faster Than Ever, NASA Says." CBS News. September 25, 2023. https://www.cbsnews.com/news/antarctica-greenland-ice-melt-nasa-climate-change/
"**mercury release**." Joy, Darrin S. "'Mercury Bomb' Threatens Millions as Arctic Temperatures Rise, Study Warns." Phys.org. August 15, 2024. https://phys.org/news/2024-08-mercury-threatens-millions-arctic-temperatures.html.
"**heat waves.**" "Climate Change: Global Temperature." Climate.gov. Accessed September 25, 2024. https://www.climate.gov/news-features/understanding-climate/climate-change-global-temperature .
"**Hurricane Beryl**." "Hurricane Beryl Breaks Climate Records." Science News. Accessed September 25, 2024. https://www.sciencenews.org/article/hurricane-beryl-climate-records.
"**Hurricane Helene**." Friedman, Lisa. "Hurricane Helene and the Rapid Intensification Trend." Axios, September 27, 2024. https://www.axios.com/2024/09/27/hurricane-helene-rapid-intensification-trend.

"**Helene's storm damage estimates**." MSN News. "Hurricane Helene Could Cost $200 Billion: Nobody Knows Where the Money Will Come From." Accessed October 17, 2024. https://www.msn.com/en-us/news/us/hurricane-helene-could-cost-200-billion-nobody-knows-where-the-money-will-come-from/ar-AA1rIxOM?form=MG0AV3
"**Flooding in Northeast**." "Intense Storms Batter Northeast, Cause Severe Flash Flooding, Submerging Roads and Homes." NBC News. Accessed September 25, 2024. https://www.nbcnews.com/news/weather/intense-storms-batter-northeast-cause-severe-flash-flooding-submerging-rcna167112 .
"**floods in the Midwest**." Associated Press. "What's Causing the Devastating Flooding in the Midwest." *U.S. News & World Report*. June 24, 2024. https://www.usnews.com/news/us/articles/2024-06-24/whats-causing-the-devastating-flooding-in-the-midwest .
"**flooding in China**." Gan, Nectar. "*Massive floods threaten tens of millions.*" CNN, April 22, 2024. URL: https://www.cnn.com/2024/04/22/china/china-guangdong-floods-intl-hnk/index.html
"**flooding in Brazil**." Al Jazeera. "'It's Going to Be Worse': Brazil Braces for More Pain Amid Record Flooding." Al Jazeera, May 4, 2024. https://www.aljazeera.com/news/2024/5/4/its-going-to-be-worse-brazil-braces-for-more-pain-amid-record-flooding .
"**U.S. Drought Monitor Map**." "U.S. Drought Monitor." https://droughtmonitor.unl.edu/ Accessed September 1, 2024.
"**Sierra Nevada snowpack**." "Sierra Nevada." Water Education Foundation. Accessed September 25, 2024. https://www.watereducation.org/aquapedia/sierra-nevada
"**wildfire impacts**." "Climate Change Indicators: Wildfires." U.S. Environmental Protection Agency. Accessed September 25, 2024. https://www.epa.gov/climate-indicators/climate-change-indicators-wildfires
"**Whale migration**." "Whales and Climate Change: Big Risks to the Ocean's Biggest Species." *NOAA Fisheries*. Accessed September 25, 2024. https://www.fisheries.noaa.gov/national/climate/whales-and-climate-change-big-risks-oceans-biggest-species
"**Forest elephants**." "Gabon's Forest Elephants Find Refuge." *National Geographic*. Accessed September 25, 2024. https://www.nationalgeographic.com/magazine/article/gabon-forest-elephants-refuge-feature
"**Giraffe populations**." "Study: Rising Rainfall, Not Temperatures, Threaten Giraffe Survival." *Phys.org*. Accessed September 25, 2024. https://phys.org/news/2023-06-rainfall-temperatures-threaten-giraffe-survival.html
"**Polar bear cubs.**" https://www.bbc.com/news/science-environment-53474445
"**1,497 endangered species**." https://insideclimatenews.org/news/03092023/research-link-between-emissions-polar-bear-decline/
"**Climate change impact on coffee**." Bilen, Christine, Daniel El Chami, Valentina Mereu, Antonio Trabucco, Serena Marras, and Donatella Spano. A Systematic Review on the Impacts of Climate

Change on Coffee Agrosystems." *Plants* 12, no. 1 (2023): 102. https://doi.org/10.3390/plants12010102 .
maize." **"Why Maize?"** MAIZE. CIMMYT. Accessed September 10, 2024. https://development.maize.org/why-maize/ .
"Australian Climate Council Infographic." https://www.climatecouncil.org.au/wp-content/uploads/2021/04/Aim-high-go-fast-infographic-Bella-Edit-scaled.jpg
Author Andrew Boyd quote." Boyd, Andrew. *I Want a Better Catastrophe: Navigating the Climate Crisis with Grief, Hope, and Gallows Humor*. New Society Publishers. 2023.
"António Guterres quote." United Nations. "Secretary-General's Opening Remarks at Press Conference on Climate." Last modified July 27, 2023. Accessed September 10, 2024. https://www.un.org/sg/en/content/sg/press-encounters/2023-07-27/secretary-generals-opening-remarks-press-conference-climate
"personal hygiene habit emissions." McDonald, Brian C., Jessica A. Gentner, Allen H. Goldstein, and Joost de Gouw. "Volatile Chemical Products Emerging as Largest Petrochemical Source of Urban Organic Emissions." Science 359, no. 6377 (2018): 760-764. https://doi.org/10.1126/science.aaq0524
"State of the Climate Action Report." Boehm, S., et al. State of Climate Action 2023. Berlin and Cologne, Germany, San Francisco, CA, and Washington, DC: Bezos Earth Fund, Climate Action Tracker, Climate Analytics, ClimateWorks Foundation, NewClimate Institute, the United Nations Climate Change High-Level Champions, and World Resources Institute, 2023. https://doi.org/10.46830/wrirpt.23.00010 .
"*Tell Them* poem excerpt." Jetñil-Kijiner, Kathy. *Poems from a Marshallese Daughter*. Tucson: University of Arizona Press, 2017. Reprinted by permission of the University of Arizona Press.
"Sea Change book." Gerhardt, Christina. *Sea Change: An Atlas of Islands in a Rising Ocean*. Berkeley: University of California Press, 2023.
"Who's Managing Your Future Report." Sierra Club. *Asset Manager Report*. June 2023. https://www.sierraclub.org/sites/www.sierraclub.org/files/2023-06/Asset_Manager_Report_V5.pdf .
"Shareholders press for disclosures." Eccles, Robert G., and Svetlana Klimenko. "*Shareholders Are Pressing for Climate Risk Disclosures. That's Good for Everyone*." Harvard Business Review, April 22, 2021. https://hbr.org/2021/04/shareholders-are-pressing-for-climate-risk-disclosures-thats-good-for-everyone .
"MoveTheMoney campaign." "Move the Money." *We Don't Have Time*. Accessed September 8, 2024. https://app.wedonthavetime.org/frontpage/movethemoney
"Green Century Fund." Sierra Club. "About Us." *Green Century*. Accessed September 26, 2024. https://www.greencentury.com/about-us/ .
"As You Sow." As You Sow. "About Us." Accessed September 26, 2024. https://www.asyousow.org/about-us .
"What does 1.5 degrees mean." Domonoske, Camila. "*What Does*

1.5 Degrees Mean in a Warming World?" NPR, November 8, 2021. https://www.npr.org/2021/11/08/1052198840/1-5-degrees-warming-climate-change .

CHAPTER FIFTEEN: AGING RAGERS

"**Hazel Chandler quote**." 19th News. "Climate Grandmothers: Environmental Activism." January 2024. https://19thnews.org/2024/01/climate-grandmothers-environmental-activism/ .
"**Full Monty quote**." Cattaneo, Peter, director. *The Full Monty*. United States: Fox Searchlight Pictures, 1997.
"**Grace & Frankie episode**." Kauffman, Marta, and Howard J. Morris, creators. *Grace & Frankie*. Season 2, episode 13, "The Coup." Directed by Rebecca Asher. Aired May 6, 2016. Netflix. https://www.netflix.com/title/80017537
"**AARP sex toys article**." https://www.aarp.org/benefits-discounts/members-only-access/info-2024/in-the-mood-buying-sex-toys.html
"**Ben Cohen quote**." Ben & Jerry's. "Ben and Jerry Arrested." Ben & Jerry's, April 18, 2016. https://www.benjerry.com/whats-new/2016/ben-and-jerry-arrested

CHAPTER SIXTEEN: OCEAN ACIDIFICATION

"**Recommended Musical Pairing**." Johnson, Jack. "Only the Ocean." Track 14 on *To the Sea*. Brushfire Records, 2010.
"**formation of calcium carbonate**." NOAA's National Ocean Service. "What is Coral Bleaching?" NOAA, June 25, 2018. https://oceanservice.noaa.gov/facts/coral_bleach.html
"**coral reefs**." Hood, Marlowe. "Flood Damage Would Double Without Coral Reefs: Study." Phys.org. June 12, 2018. https://phys.org/news/2018-06-coral-reefs.html
"**importance of coral reefs**." NOAA's National Ocean Service. "Why Are Coral Reefs Important?" NOAA, June 25, 2018. https://oceanservice.noaa.gov/education/tutorial_corals/coral07_importance.html
"**phytoplankton and oxygen**." Green.org. "Aqualung: Why You Should Really, Really Care About Phytoplankton." Green.org, September 30, 2019. https://green.org/2019/09/30/aqualung-why-you-should-really-really-care-about-phytoplankton
"**why care about phytoplankton quote**." Ibid.
"**polymetallic nodules**." Perkins, Sid. "In a Seafloor Surprise, Metal-Rich Chunks May Generate Deep-Sea Oxygen." *Science News*, July 22, 2024. https://www.sciencenews.org/article/seafloor-metal-nodules-deep-sea-oxygen
"**Clarion-Clipperton Zone critters**." "Underwater Creatures: 5,000 New Species Found in the Pacific Ocean." *BBC News*, May 26, 2023. https://www.bbc.com/news/science-environment-65722878

"**Deep Sea Conservation Coalition.**" "Governments and Parliamentarians." *Deep Sea Conservation Coalition*. Accessed September 11, 2024. https://deep-sea-conservation.org/solutions/no-deep-sea-mining/momentum-for-a-moratorium/governments-and-parliamentarians/
"**salt batteries**" and "**mining nodules.**" **John Oliver Deep Sea Mining episode**." Last Week Tonight. "Deep-Sea Mining: Last Week Tonight with John Oliver." *YouTube*, June 13, 2024. 23:18. https://youtu.be/qW7CGTK-1vA?si=tAIQBDX_Zqkxm7g9
"**Chinese automaker using salt batteries.**" "China's Top EV Battery Maker CATL to Supply Honda." *Yahoo Finance*, accessed September 11, 2024. https://finance.yahoo.com/news/chinas-top-ev-battery-maker-082125820.html
"**companies pledge not to use nodules.**" David Shukman, "Deep Sea Mining: BMW, Volvo, Google and Samsung Call for Moratorium," *BBC News*, April 3, 2021. https://www.bbc.com/news/science-environment-56607700
LADWP data." "Facts & Figures." *Los Angeles Department of Water and Power*. Accessed September 11, 2024. https://www.ladwp.com/who-we-are/power-system/facts-figures
Citizen Science. https://www.citizenscience.gov/
NeMO-Net Video Game. http://nemonet.info/
NASA Science. https://science.nasa.gov/citizen-science/
iNaturalist. https://www.inaturalist.org/
Society for Science. https://www.societyforscience.org/research-at-home/citizen-science/

CHAPTER SEVENTEEN: FROM THE STANDUP STAGE

"**Lincoln and humor**." Sachs, Aaron. Stay Cool: Why Dark Comedy Matters in the Fight Against Climate Change. New York University Press. 2023.

CHAPTER EIGHTEEN: AIR POLLUTION

"**Recommended Musical Pairing**." Lehrer, Tom. "Pollution." Track 7 on *That Was The Year That Was*. Reprise Records, 1965.
"**brown cloud**." Ahmad, K. "Pollution Cloud Over South Asia Is Increasing Ill Health." *The Lancet* 360, no. 9341 (2002): 1261. https://doi.org/10.1016/S0140-6736(02)11367-7.
"**Clean Cooking Alliance**." https://cleancooking.org/
"**Modern Cooking Alliance for Africa.**" https://www.moderncooking.africa/
"**Solar Cookers International.**" https://www.solarcookers.org/
"**Kamala Harris quote**." Harris, Kamala. "Remarks by Vice President Kamala Harris at the Celebration of America." The White House. January 20, 2021. https://www.whitehouse.gov/briefing-room/speeches-remarks/2021/01/20/remarks-by-vice-president-kamala-harris-at-the-celebration-of-america

CHAPTER TWENTY: OZONE

"**Recommended Musical Pairing**." Sesame Street. "That's Cooperation." Performed by Big Bird and the birds. Music by Michael Kosarin, lyrics by Molly Boylan. First aired in Episode 4406. Sesame Street Inc., 2013.
"**Kofi Annan quote**." **U.S. Department of State.** "The Montreal Protocol on Substances That Deplete the Ozone Layer." Accessed September 12, 2024. https://2009-2017.state.gov/e/oes/eqt/chemicalpollution/83007.htm
"**Susan Strahan quote**." Merzdorf, Jessica. "NASA Data Aids Ozone Hole's Journey to Recovery." NASA, September 16, 2020. https://www.nasa.gov/earth-and-climate/nasa-data-aids-ozone-holes-journey-to-recovery/ .

CHAPTER TWENTY-ONE: I GET IT

"**Chesterfield slogan**." Dach, Jeffrey. "Doctor Says Trust Me, Cigarettes Are Healthy." Jeffrey Dach MD, 2013. Accessed September 12, 2024. https://jeffreydachmd.com/doctor-says-trust-me-cigarettes-are-healthy
"**overfishing**." United Nations Framework Convention on Climate Change (UNFCCC). "Plenty of Fish." Accessed September 26, 2024. https://unfccc.int/news/plenty-of-fish .
"**palm oil.**" Save the Orangutan. "What Is RSPO?" Accessed September 26, 2024. https://savetheorangutan.org/what-is-rspo/ .
"**Land Cruiser emissions**." U.S. Department of Energy. "2024 Toyota Land Cruiser." *FuelEconomy.gov*. Accessed September 26, 2024.
https://www.fueleconomy.gov/feg/Find.do?action=sbs&id=47762 .
"**Commercial aviation emissions**." Graver, Brandon, Kevin Zhang, and Dan Rutherford. *CO2 Emissions from Commercial Aviation: 2013, 2018, and 2019*. International Council on Clean Transportation, October 2020.
https://theicct.org/sites/default/files/publications/CO2-commercial-aviation-oct2020.pdf .
"**Environmental costs of fast fashion**." United Nations Environment Programme (UNEP). "Environmental Costs of Fast Fashion." November 12, 2018. https://www.unep.org/news-and-stories/story/environmental-costs-fast-fashion .
"**Clothing purchasing**." Balchandani, Anita, Marco Beltrami, Achim Berg, Saskia Hedrich, Felix Rölkens, and Imran Amed. "The End of Ownership for Fashion Products." McKinsey & Company. Accessed September 12, 2024. https://www.mckinsey.com/industries/retail/our-insights/the-end-of-ownership-for-fashion-products
"**Conscious Chatter**." Conscious Chatter. "Where What We Wear Matters." *Conscious Chatter*. Accessed September 12, 2024. https://consciouschatter.com/"
"**Laura François**." https://www.awe.exchange/

CHAPTER TWENTY-TWO: PLAYLIST BONUS TRACK

"**Recommended Musical Pairing**." Shanks, John, Tobias Gad, Sean Henriques, and Natasha Bedingfield. "Love Song to the Earth." Gad Music, 2015.
"**Buy Nothing**." https://buynothingproject.org/find-a-group
"**Freecycle.**" https://www.freecycle.org/
"**surprising things to borrow from library.**" Annie Suhy, "75 Surprising Things You Can Borrow from the Library," *Libby Life*, July 25, 2024, https://www.libbylife.com/2024-07-25-75-surprising-things-you-can-borrow-from-the-library
"**Find a time bank.**" https://hourworld.org/index.htm
"**The Time Bank Solution**." https://ssir.org/articles/entry/the_time_bank_solution
"**Repair Cafes.**" https://www.repaircafe.org/en/

CHAPTER TWENTY-THREE: CRIPPLING IGNORANCE

"**erased climate change**." Luscombe, Richard. "Climate Deniers like DeSantis Hurt Most Vulnerable Communities, Scientists Say." TheGuardian.Com. June 1, 2024. https://doi.org/https://www.theguardian.com/environment/article/2024/jun/01/ron-desantis-climate-deniers-hurt-communities
"**record temperatures**." Borges, Carolina, Alex Browning, Vanessa Medina, and Rubén Rosario. "South Florida Faces Record Spring Temperatures; Health Officials Urge Caution." WSVN News. May 15, 2024. https://wsvn.com/news/local/miami-dade/south-florida-faces-record-spring-temperatures-health-officials-urge-caution/
"**Florida blocks worker protections**." **Tallahassee Democrat.** "Ron DeSantis Signs Bill Blocking Heat Protections for Florida Workers." *Tallahassee Democrat*, April 15, 2024. https://www.tallahassee.com/story/news/2024/04/15/ron-desantis-bill-blocking-heat-protections-florida-workers/73324894007/
"**Texas blocks water breaks**." Mark Felix. "Backlash Brews as Texas Law Eliminates Mandatory Water Breaks for Construction Workers." NBC News, August 10, 2023. https://www.nbcnews.com/science/science-news/backlash-brews-texas-law-eliminates-mandatory-water-breaks-rcna92961
"**videos deny climate science.**" Milman, Oliver. "Videos Denying Climate Science Approved by Florida as State Curriculum." The Guardian. August 10, 2024. https://www.theguardian.com/us-news/2023/aug/10/florida-ron-desantis-climate-vidoes-school-curriculum
"**Insurify report.**" Insurify. "The Most and Least Climate-Resilient Cities for Homeowners." *InsuranceNewsNet*, September 13, 2024. https://insurancenewsnet.com/oarticle/the-most-and-least-climate-resilient-cities-for-homeowners-insurify
"**fossil fuel companies fund college institutes**." "The Fossil Fuel

Industry's Invisible Colonization of Academia." *The Guardian,* March 13, 2017. https://www.theguardian.com/environment/climate-consensus-97-per-cent/2017/mar/13/the-fossil-fuel-industrys-invisible-colonization-of-academia?ref=drilled.ghost.io .
"**Katharine Hayhoe quote**." Hayhoe, Katharine. *Saving Us: A Climate Scientist's Case for Hope and Healing in a Divided World.* New York: Atria/One Signal Publishers, 2021. Page 51.
"**Sesame Street climate programming**." Hiltzik, Michael. "How 'Sesame Street' Can Prepare Kids for Climate Disasters." *Los Angeles Times,* May 23, 2024. https://www.latimes.com/environment/newsletter/2024-05-23/column-how-sesame-street-can-prepare-kids-for-climate-disasters-boiling-point
"**Sesame Street impact**." Sesame Workshop. *Sesame Workshop International Impact Report.* Accessed September 26, 2024. https://downloads.cdn.sesame.org/sw/SWorg/documents/SW_international_impact.pdf .
"**why attend a school board meeting**." Terri Huggins Hart. "Very Good Reasons to Attend a School Board Meeting." *Parents,* accessed September 13, 2024. https://www.parents.com/parenting/better-parenting/very-good-reasons-to-attend-a-school-board-meeting/
"**time in nature**." DeVille, Nicole V et al. "Time Spent in Nature Is Associated with Increased Pro-Environmental Attitudes and Behaviors. International Journal of Environmental Research and Public Health. July 2021. https://www.ncbi.nlm.nih.gov/pmc/articles/PMC8305895
"**Beyond a Book**." https://www.beyondabook.org/

CHAPTER TWENTY-FIVE: RANTING IN THE STREET

Chödrön, Pema. *When Things Fall Apart: Heart Advice for Difficult Times.* Boston: Shambhala Publications, 1997.

CHAPTER TWENTY-SIX: ENVIRON-MENTAL

"**Jenny Price quote**." Price, Jenny. Stop Saving the Planet!: An Environmentalist Manifesto (p. 23). W. W. Norton & Company. Kindle Edition.
"**ExxonMobil predicted climate crisis quote**." Carly Cassella. ExxonMobil Predicted the Climate Crisis 5 Decades Ago, Leaks Show." *ScienceAlert.* Accessed September 13, 2024. https://www.sciencealert.com/exxonmobil-predicted-the-climate-crisis-5-decades-ago-leaks-show .
"**ExxonMobil op-ads.**" Westervelt, Amy. *Drilled,* season 1, "The Origins of Climate Denial." Podcast audio. 2018. https://drilled.media/podcasts/drilled
"**Supran and Oreskes quote**." Supran, Geoffrey, and Naomi Oreskes. "Yes, ExxonMobil Misled the Public about Climate Change." *The New York Times,* August 22, 2017.

https://www.nytimes.com/2017/08/22/opinion/exxon-climate-change-.html
"**DeSantis signs legislation**." Office of Governor Ron DeSantis. "Governor Ron DeSantis Signs Bill to Further Strengthen Florida's Resiliency Efforts." *Florida Governor's Office*, May 12, 2021. https://www.flgov.com/2021/05/12/governor-ron-desantis-signs-bill-to-further-strengthen-floridas-resiliency-efforts/
"**new jobs in clean energy**." U.S. Department of Energy. "DOE Report Finds Clean Energy Jobs Grew in Every State in 2022." Last modified September 13, 2024. https://www.energy.gov/articles/doe-report-finds-clean-energy-jobs-grew-every-state-2022
"**Texas passed anti-renewable energy laws**." Metzger, Luke. "How the Environment Fared at the Texas Legislature (88th Regular Session)." Environmentamerica.Org. May 29, 2023. https://environmentamerica.org/texas/articles/how-the-environment-fared-at-the-texas-legislature-88th-regular-session/
"**Rachel Carson quote**." Carson, Rachel. Silent Spring. Boston: Houghton Mifflin, 1962.
"**no friends across the aisle**." John Gramlich. "20 Striking Findings From 2020." *Pew Research Center*, December 11, 2020. https://www.pewresearch.org/short-reads/2020/12/11/20-striking-findings-from-2020/
"**pipeline impacts**." See the See the Mountain Media Productions YouTube channel for video documentation. https://youtube.com/@oliverphotographygreenville?si=JBRouZlVoIOPmCMR
"**An Unreasonable Woma**n." Wilson, Diane. *An Unreasonable Woman: A True Story of Shrimpers, Politicos, Polluters, and the Fight for Seadrift, Texas*. White River Junction, VT: Chelsea Green Publishing, 2005.
"**Molly Ivans quote**." https://dianewilsonactivist.org/about/

CHAPTER TWENTY-EIGHT: HERE TO STAY

"**Rob Brezsny quote**." Brezsny, Rob. *Pronoia Is the Antidote for Paranoia: How the Whole World Is Conspiring to Shower You with Blessings*. Berkeley, CA: North Atlantic Books, 2005.

CHAPTER TWENTY-NINE: CLOSING THOUGHTS

"**dreamer's disease song**." New Radicals. "You Get What You Give." *Maybe You've Been Brainwashed Too*. MCA, 1998
"**definition of disease**." Merriam-Webster, "Word History of Disease," Merriam-Webster, accessed September 16, 2024, https://www.merriam-webster.com/wordplay/word-history-of-disease .
"**Holy Universe quote**." Christopher, David. The Holy Universe: A New Story of Creation for the Heart, Soul, and Spirit. New Story Press. 2014.

"**numbers too big to ignore song**." Reddy, Helen. "I Am Woman." Capitol Records, 1972. Accessed August 29, 2024. https://lyrics.com/ .
"**25% needed for change**." Centola, Damon, Joshua Becker, Devon Brackbill, and Andrea Baronchelli. 2018. "Experimental Evidence for Tipping Points in Social Convention." Science 360 (6393): 1116-1119. https://www.science.org/doi/10.1126/science.aas8827 .
"**Love Her Madly song**." The Doors. "Love Her Madly." Elektra Records, 1971. Accessed August 29, 2024. https://lyrics.com/ .
"**Changemaker quiz**." The Story of Stuff Project. "What Kind of Changemaker Are You?" Accessed September 12, 2024. https://action.storyofstuff.org/survey/changemaker-quiz/
"**Chaotic climate book**." Schmidt, LaUra with Aimee Lewis Reau & Chelsie Rivera. How to Live in a Chaotic Climate: 10 Steps to Reconnect with Ourselves, Our Communities and Our Planet. Shambala Publications. 2023.

CHAPTER THIRTY-ONE: GOING FORTH

"**Alice Walker quote**." Alice Walker, Hard Times Require Furious Dancing. New World Library. 2010.

RESOURCES

BOOKS
Attenborough, David. *A Life on Our Planet: My Witness Statement and a Vision for the Future*. New York, NY: Grand Central Publishing, 2020.
Boyd, Andrew. *I Want a Better Catastrophe*. New Society Publishers. 2023.
Cole, Lily. *Who Cares Wins: Reasons for Optimism in Our Changing World*. London: Penguin Books, 2020.
Gates, Bill. *How to Avoid a Climate Disaster: The Solutions We Have and the Breakthroughs We Need*. New York: Alfred A. Knopf, 2021
Macy, Joanna, and Molly Brown. *Coming Back to Life: The Updated Guide to the Work That Reconnects*. Gabriola Island, BC: New Society Publishers, 2014.
Stoknes, Per Espen. *What We Think About When We Try Not To Think About Global Warming: Toward a New Psychology of Climate Action*. White River Junction, VT: Chelsea Green Publishing, 2015.

PODCASTS
Sustainable Minimalist podcast. https://mamaminimalist.com/category/podcast/
Outrage & Optimism podcast. https://www.outrageandoptimism.org/
EcoJustice Radio podcast. https://www.youtube.com/@ecojusticeradio

Release

This publication is designed to provide accurate information in regard to the subject matter covered at the time it was authored. It is sold with the understanding that neither the author nor the publisher is engaged in rendering legal, investment, accounting or other professional services. While the publisher and author have used their best efforts in preparing this book, they make no representations or warranties with respect to the accuracy or completeness of the contents of this book and specifically disclaim any implied warranties of merchantability or fitness for a particular purpose. No warranty may be created or extended by sales representatives or written sales materials. The advice and strategies contained herein may not be suitable for your situation. You should consult with a professional when appropriate. Neither the publisher nor the author shall be liable for any loss of profit or any other commercial damages, including but not limited to special, incidental, consequential, personal, or other damages.

Made in the USA
Columbia, SC
23 November 2024